Leadership Basics for Frontline Managers

Bill Templeman

Leadership Basics for Frontline Managers

Tips for Raising Your Level of Effectiveness and Communication

CRC Press
Taylor & Francis Group
Boca Raton   London   New York

CRC Press is an imprint of the
Taylor & Francis Group, an **informa** business

A PRODUCTIVITY PRESS BOOK

CRC Press
Taylor & Francis Group
6000 Broken Sound Parkway NW, Suite 300
Boca Raton, FL 33487-2742

© 2014 by Bill Templeman
CRC Press is an imprint of Taylor & Francis Group, an Informa business

No claim to original U.S. Government works

Printed on acid-free paper
Version Date: 20140210

International Standard Book Number-13: 978-1-4822-1995-1 (Paperback)

Visit the Taylor & Francis Web site at
http://www.taylorandfrancis.com

and the CRC Press Web site at
http://www.crcpress.com

Contents

SECTION III   Communication

SECTION IV   Your Career

Preface

THE CHAPTERS

When I started writing the column in which these chapters first appeared as articles, I asked my editor to select a topic from a list of ideas I would submit prior to each deadline. As we became more used to working together, she eventually turned over the job of topic selection to me. All of which is to say that most of these topics have grown out of my own business experience as an independent consultant. There is much more that could be said about every one of these topics, but I decided it was better to focus on the essential learning points, and to make them as practical as possible. Each chapter ends with a list of bullet points or actions the reader might take to address a given situation. Consider this collection of articles to be a toolkit; reach in and take what you need and when you need it.

Note: I have taken appropriate liberties in fictionalizing scenarios to support the learning points and protect client confidentiality. A few of the chapters are responses to books or articles I have come across in my work. I have listed sources wherever appropriate.

Acknowledgments

While many colleagues and clients deserve to be acknowledged here for the ideas and shared experiences that led to this book, I need to offer my thanks, in particular, to a few people without whom the creation of this book would not have been possible.

First of all, I wish to thank Ken Victor, a friend and colleague at the Edgework Leadership Group, for his steadfast encouragement and inspired feedback.

Thanks also to Jeff Macklin of Prevail Media for his design expertise and assistance in preparing this book for publication.

Finally, thanks to my partner, Trudi Ruch, and my daughters, Emily and Hannah, who created the essential space and time in order for this work to proceed.

Introduction

HOW TO READ THIS BOOK

Most of the management skills chapters (25 in all) selected for inclusion in this book began life in 2006 as columns in the *Durham Trade & Commerce Magazine* and, since January 2011, in the *Durham Business Times*. I have grouped these chapters under three general headings: Personal Effectiveness, Leadership, and Communication. This introduction explores the origins of the chapters, what they are trying to achieve, and the context in which they were written.

In writing these chapters, I have tried to speak directly to that most underserved segment of the contemporary workforce: the frontline manager. These are the people who, as often as not, are promoted from operational jobs into positions that require them to manage not only work processes, but the people who do the work. They sometimes arrive in their new positions without sufficient formal managerial training in those most ill-defined of managerial competencies, "people skills." As a program designer and facilitator, my rewards are based on the levels of engagement participants display while in seminars. When I see enthusiastic involvement during a seminar and later receive appreciative feedback, I know that I have tapped into their neglected desire for development.

Why is this desire for development being neglected by organizations? Spending restraint is the easy answer, but not the only one. In an era of strategic management and high executive control, investment in employee development, particularly when imposed from above without a detailed business case, without feedback from the intended audience, without a needs analysis or implementation strategy, and without strategic alignment to an organization's business plan, can indeed lead to negative returns: loss of time, loss of money, and, more importantly, incalculable losses in trust and credibility.

However, it doesn't have to turn out this way. Properly done, good employee development can improve careers and unleash the power of organizations to achieve great things. I make no apologies for being messianic on this point: Good training can change lives.

How can organizations "get it right" in terms of how they train their frontline managers?

Too often, training in management skills, particularly on communication, either does not happen at all, or happens sporadically in a patchwork fashion that has no focus or tie-in to the rest of the organization. Training budgets, if they still exist, are viewed as being thoroughly dispensable when competing business priorities collide. In large organizations, the tendency is to continue training, but, in order to manage tight budgets, there is a trend toward doing all training individually and online, often without follow-up. Online training, applied appropriately, has immense potential for improving performance. However, a webinar, just like a classroom seminar, is only a delivery vehicle for learning activities. Regardless of the vehicle, **the learning needs to be aligned, reinforced, and modeled throughout the hosting organization** in order for a decent return on investment to be realized.

Most unfortunate of all is the slippage in assumed common knowledge that has happened over the past decade. What was common knowledge 10 years ago and unnecessary to repeat now requires detailed explanation. Not so long ago I had to launch into a detailed explanation of SMART (specific, measurable, attainable, relevant, and time-bound) objectives. SMART objectives were as common as hammers and nails during the Quality Movement of the 1980s. Not anymore. **Organizations can no longer assume a common knowledge base.**

Digital technology is changing our learning styles. Lectures and texts are becoming increasingly irrelevant as useful instructional tools. Attention spans are shrinking. Tolerance for reading is shrinking. Part of getting training right means that **learning has to be parceled into very time-efficient packages in order to maximize uptake.**

The complexity of work is accelerating at an explosive rate. Employees need to know much more today in order to do the jobs expected of them. More than that, the culture of work is being stretched and compressed by opposing trends at the same time. We have less time for the human side of work. Spans of control have widened significantly over the past two decades. **The success of any training initiative depends on the acceptance of the reality that we must all work through multiple paradoxes every day.**

Training is being downgraded as a business improvement tactic at exactly the wrong time. Jobs have become much more complex thanks to technology and the evolution of how we do business. There is much, much more information available today for a supervisor on how to do her or his job compared to 30 years ago. But there is much less time available to learn how to do that job. So, the paradox becomes one of increased wealth and increased poverty; there is an infinite amount of information available on how to work, and an ever-diminishing amount of time in which to delve into this deep mine of information. We need more knowledgeable frontline managers at precisely the moment in organizational life when the time needed for knowledge acquisition and skill building is not given sufficient legitimacy. **Getting it right means that learning has to be extremely time-efficient, focused, and based on real-time, present-tense needs. At the same time, learning needs to generate agility and resilience to respond to the unknown and unpredictable. We have to develop not only skills and knowledge, but mindset and attitude.** This collection of chapters is meant to be of service in "getting it right."

Section I

Personal Effectiveness

1

Extreme Organizational Politics: Wishfulness and Yesmanship

Imagine what your work would be like if the fear bred by organizational politics did not exist. Imagine workplaces in which employees were encouraged to tell the truth, no matter how unpalatable this truth might be for their executives. Imagine what coming to work every day would be like if integrity was the primary operating principle.

Any organization that is dominated by extreme levels of organizational politics can become toxic and fail. When I worked for a company in the financial industry, I was hyperconcerned about conforming and getting ahead. I wore the right clothes, worked the right amount of hours, contributed during meetings in appropriate ways, and, in general, did my best to fit in, to be highly valued, and to get promoted. Or so I thought. When my firm began to take imprudent business risks, the rumor mill ran wild. I remember being told by a colleague that someone in accounting had said, "You just wouldn't believe what the executive team is telling us to do with the books!" The numbers were being manipulated to hide the true picture from shareholders, yet very few of us were willing to become whistle blowers. Very few of us dared to challenge the directions that were coming down from the executive suite. The company's share price gradually slid from over $20 to below 50¢. Terminations became the order of the day. Eventually the company was sold.

The corrosive fear that undermines organizational success has a long history. Writing shortly after the end of World War II, Admiral John Godfrey, the former director of Britain's Naval Intelligence Division, in analyzing Operation Mincemeat, a highly successful wartime deception conducted by British agents, identified two major weaknesses of the Nazi's espionage establishment: "wishfulness" and "yesmanship." These words are strictly the good admiral's concoctions. Yet wishfulness and yesmanship have changed the course of history, and they are still with us today, every day of the week, at work and at home.

As definitions for wishfulness and yesmanship do not appear in any dictionary, I'll use my own. *Wishfulness* is that tendency among individuals and organizations to believe information that supports their own view of reality while simultaneously rejecting all contradictory information. Godfrey believed that the Nazi high command, when presented with two pieces of contradictory information, was "inclined to believe the one that fit in best with their own previously formed conceptions."

Yesmanship is the tendency of those with less positional power to agree with those who have greater power, mainly out of fear. Yesmanship feeds on fear of authority; the greater the fear, the stronger the tendency toward blind yesmanship. Yesmanship is an enabling behavior for wishfulness. Wishfulness, particularly in organizations in which there are dire consequences for insubordination, can give rise to deadly levels of yesmanship.

Milder forms of yesmanship occasionally take a seat at almost every corporate or government boardroom. Fearful employees learn instinctively to deliver the news they believe their harried bosses want to hear. Wishing to avoid an argument, employees will spin information for each other by hiding in yesmanship: "Don't make waves." "Tell her what she wants to hear and you'll be fine."

In the Nazi military command hierarchy that Godfrey analyzed, lower ranked personnel would deliberately distort information in order to crawl higher in Hitler's estimation. Yesmanship became integrated into strategic decision making at the highest levels of the Third Reich. In this rigidly hierarchical military structure, no one dared say "no" to the powers above. Wishfulness and yesmanship ultimately destroyed the Nazi war machine.

What power does wishfulness and yesmanship have in your organization today? Who could give you an honest answer?

We all know what organizational politics can feel like. We all know the almost imperceptible sense of caution, of carefulness, of not wanting to communicate the wrong message. We all know the importance of maintaining a professional image, or being seen to be a worthwhile contributor, of being perceived as someone who supports current directions and plans. All of which is not to say that by being careful, considerate, conscious of one's image and messages we are somehow sabotaging our careers. Far from it. However, it is a question of degree. How careful do we need to be?

We all know the cost of not being careful enough, but, do we understand the cost of being too careful?

The costs of allowing these forms of organizational cowardice to become the norm could be immense. What can we do to ensure that wishfulness and yesmanship do not distort our business planning and operational decision making? How can we encourage people to speak their truth? How can we build an organizational culture of high integrity? (Figure 1.1)

- Everyone, from the CEO on down, to paraphrase Mahatma Gandhi, must be the change they want to see in their colleagues. If you want the truth, you must speak the truth and be the truth.
- Encourage debate and dialogue. Welcome challenges, welcome questions, welcome demands for explanations, and, above all, welcome alternative ideas that conflict with your own assumptions.
- If you are a leader, make a point of hiring people who are likely to disagree with you on business issues. Conflict can yield creative resolutions that would never see the light of day had passive politeness been the name of the game.
- Instead of arguing with dissenters, ask for explanations of their thinking. "How did you come to that conclusion? Please walk me through your thinking process." Listen first before fighting back.
- Treat everyone according to a set of explicit and worthwhile values. This is not about posting flowery vision statements everywhere. This is about your behavior, or more accurately, how you treat people, all people, every day.
- Build a culture of trust by demonstrating trust. You must believe in the people with whom you work. You believe that they have the best of intentions and that they fully deserve your trust.
- Show your commitment by demonstrating everything you believe in through your own behavior.
- Developing a culture free of wishfulness and yesmanship does not depend on the oratorical skills of a Barack Obama. Developing such a culture depends on being conscious at every moment of the messages you are sending through your every action. People learn much more about you as a leader by watching your actions as opposed to listening to you or reading your words.

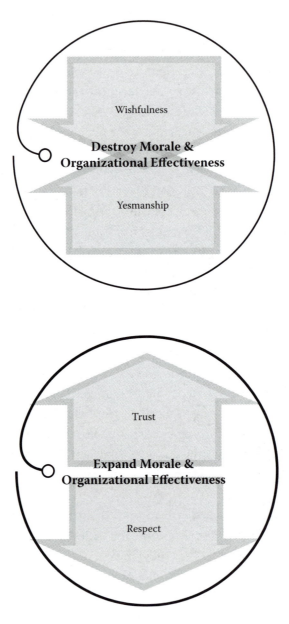

FIGURE 1.1
Wishfulness and Yesmanship vs. Trust and Respect.

- Beware of *Groupthink*, that creeping sycophantism wherein people try too hard to fit in. Cherish your dissenters as well as critiques. At times they may be frustrating to deal with, but you can always count on them to speak their truth.

Remember that conflict can, up to a point, be a sign of organizational vitality. If people really care about their work, conflicts will sometimes happen. By all means, do your best to resolve these conflicts, but don't prevent them from springing to life. They could be a healthy sign.

HOW TO FIGHT EXTREME ORGANIZATIONAL POLITICS

- AS A LEADER, MODEL THE CHANGE you want to see in others.
- ENCOURAGE DEBATE and disagreement; cherish dissenters.
- EMBRACE CONFLICT; ask dissenters why they disagree with you.
- TREAT EVERYONE with an explicit and worthwhile set of values every day.
- BUILD TRUST by demonstrating trust.
- BE SURE to practice what you are preaching.
- BE CONSCIOUS of the messages you are sending through your actions.

2

How to Work with Information Overload

No matter what we do for a living, all of us face an avalanche of distractions each day that can throw us off our game. We must constantly keep asking ourselves: "What do I have to pay attention to right now, what can wait until later, what might be good to know but not essential to my success, and what can I safely ignore?"

According to a January 2011 article in the *McKinsey Quarterly* (http://www.mckinsey.com/insights/organization/recovering_from_information_overload), the challenge of staying focused in the midst of an excess of information predates the Computer Age.

Writing in 1967, management guru Peter Drucker recommended that executives reserve large blocks of time on their calendars for thinking, not answer the phone, and return calls only once or twice a day. While Drucker's readers in the 1960s didn't have to deal with digital technology, his admonitions nonetheless still ring true today.

Those with leadership responsibility at any level face a torrent of email messages, phone calls, text messages, Tweets, Facebook postings, blog comments, and messages on other social media platforms that might contain useful customer feedback or information about competitors, new products, and business trends. How can we avoid being buried by this information tsunami?

Many of us believe that if we excel at multitasking, we can stay on top of this wave. Research cited in the McKinsey article reveals that multitasking is an ineffective coping mechanism leading to lower productivity, lower creativity, and impaired decision making. The reality is that multitasking slows us down. The human brain does its best work when focused on one task at a time. Individuals may dispute this conclusion, but the evidence is in. For example, why is texting now illegal while driving?

The inefficiency caused by multitasking is due to the brain's inability to let us perform two actions at the same time. While multitasking may allow us to cross off simpler tasks on our to-do lists, it rarely helps us resolve more difficult problems. Multitasking can become simple procrastination.

Multitasking also can make us anxious. People required to multitask show higher levels of stress. The information overload associated with multitasking lessens job satisfaction and can disrupt personal relationships. And, multitasking can become addictive by causing specific "emergency" hormones (e.g., adrenaline) to be released in our bodies.

So, if multitasking doesn't work, what can we do? (Figure 2.1)

- Be highly disciplined in how you use your time.
- Constantly set and update your priorities.
- Be focused on what matters most. Beware of cruising through information that may be nice to know, but not essential for the tasks at hand. As one CEO said, "You have to guard against the danger of overeating at an interesting intellectual buffet."
- Encourage your colleagues to respect your priorities. One colleague of mine used to place a "Focus Time" sign at her workstation. She made it clear to all of us that unless the business was in immediate jeopardy and her input critical to resolving a crisis, we were to leave her alone. She only posted this sign a few times a week, and usually only for a few hours, so we respected her wishes.
- Filter the information as it comes at you. Know what you can ignore, what you can skim, what you must read in detail later, and what you must deal with right now.
- Give your brain downtime during the workday to solve problems and reset your priorities so that you are focusing on the right things. A quick walk, a short workout, and a set period of time away from all communication technology can all help the brain to do its best work.

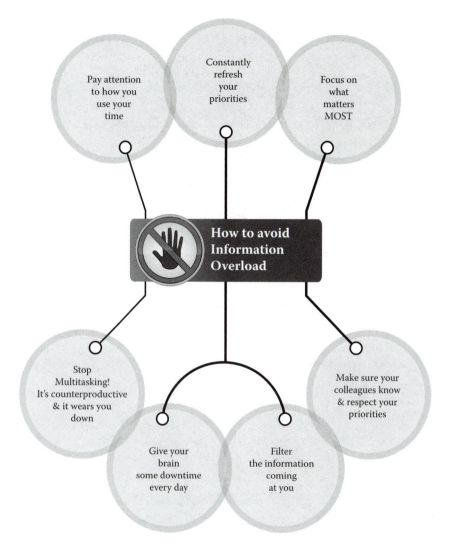

FIGURE 2.1
7 Ways to Avoid Information Overload.

- Accept the fact that multitasking is not heroic. Understand that it is really counterproductive. Instead of doing a half-baked job on five tasks at once, then be forced to take additional time to fix your mistakes, work on one task at a time, but get it right the first time.

HOW TO AVOID INFORMATION OVERLOAD

- PAY ATTENTION to how you use your time.
- CONSTANTLY REFRESH your priorities.
- FOCUS on what matters most.
- Make sure your colleagues know and RESPECT YOUR PRIORITIES.
- FILTER THE INFORMATION coming at you.
- GIVE YOUR BRAIN SOME DOWNTIME every day.
- STOP MULTITASKING. It's counterproductive and it wears you down.

3

Asking Better Questions

When we run into problems in our businesses or in our careers, we assume that we need solutions and the quicker the better. When we reach a solution, we like to believe that we have the answer to our problem. Case closed. Now move on to the next challenge.

We live in a culture that demands immediate answers. No matter what your business or what industry or sector you work in, you can be sure that there are many consultants, pundits, and instant experts just itching for your call in order to dispense their advice, sometimes for outrageously big bucks. "Solutions for sale." The business press and the consulting industry offer a myriad solutions, some valid, others less so, for every imaginable work or career problem.

Of course, answers to questions make the world go around. If you were unhappy with your business or dissatisfied in your job, you probably would want answers to your questions. Should you act on the first answer that pops up? Or, are you even ready to start looking for answers? What you really need to ask yourself is: "Am I asking the right question?"

Let's take an example. Sue, a customer service representative for a large financial services company, feels that she has no options for advancement at her current employer. She has been with the firm for eight years, is in her early 30s, and has a community college diploma in business administration. She wants to become a manager. She is quiet, methodical, and skilled at dealing with very difficult customers. She is at her best when working on her own. Her performance reviews are good, but not great. She wants higher pay. She feels that every supervisor she has ever had has stereotyped her: "Solid contributor, dependable and thorough, but not a star. No leadership potential."

Sue could frame her career question as: "How can I find another job with an organization that will help me become a manager?"

What's wrong with this question?

The question assumes that Sue really knows where she is going and how she wants to get there. The question also assumes that Sue has an image in her mind of what she wants to do, who she wants to be as a professional, the contributions she wants to make, the values that are important to her, and what sort of organization in which she would like to work.

Most of us are pretty good at challenging others' assumptions. Sometimes we are not so good at challenging our own assumptions. We are not so good at pushing ourselves to come up with the better question. This is where a coach can help.

Through coaching, Sue might discover that she has made a number of assumptions that merit closer examination. For a start, apart from the better pay, why does she want to be a manager? We know she can be a solid contributor and is dependable, but what are her dreams, her talents, her strengths, and her passions? What are the achievements she would like to look back on with pride at the end of her career? What is most important to her in any job? What risks might she be reluctant to take in order to achieve her goals?

A coach could not, and should not, give Sue any specific career advice. However, a coach might help her recognize her talents and help her get in touch with work that she is good at and that she really enjoys. By helping her to come up with better questions, a coach could help Sue get much closer to the core of what she really wants to do with her career. It would be impossible to say where her journey of self-discovery would take her, but it would be entirely possible that she would come out of the coaching process—maybe after a few sessions, maybe after a few months—with a much stronger sense of where she is going and what she wants to do.

People who are highly self-actualized in their careers—those who live their dreams and achieve their goals—are much happier in their work. This deep level of joy and satisfaction can frequently lead to greater recognition from colleagues and employers. It is no accident that people who are at the top of their games in their jobs (widely acknowledged as being top performers and great role models no matter what their level or job title) really enjoy their work and frequently earn more money than their peers who merely hold down their jobs, do what they are told, show up on time, but never really catch on fire.

So, perhaps Sue will quit the bank, go back to college, and become a paralegal and eventually become a valued member of a law firm working on human rights issues. Perhaps she will find a position in healthcare where she can excel in what she does best, which is helping frightened people deal with difficult decisions. Perhaps she will become a career counselor with homeless kids and eventually run her own placement agency. Or, perhaps she will stay at the bank and find a new position in estate administration where her caring and her unique client relationship skills will win her the trust of hundreds of clients, to say nothing of substantial bonuses.

Her coach could not have prescribed or predicted any of these outcomes. Why? Because the motivation would have to come from within as she unleashed her own inner power of self-discovery.

Sue would deserve all the credit for taking charge of her career and fulfilling her own potential. By pushing herself to ask better questions (Figure 3.1), she might discover that the keys to her own career fulfillment lie almost entirely within her control.

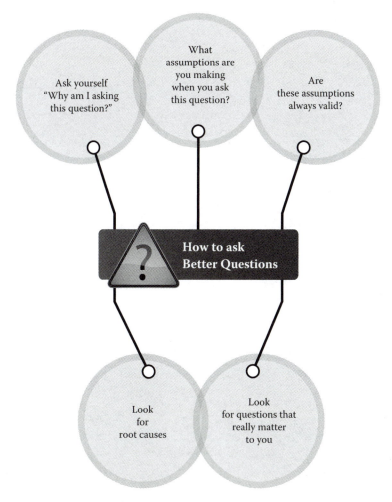

FIGURE 3.1
5 Ways to Ask Better Questions.

ASKING BETTER QUESTIONS

- ASK YOURSELF: "Why am I asking this question?"
- WHAT ASSUMPTIONS are you making when you ask your question?
- Are these ASSUMPTIONS ALWAYS VALID?
- LOOK for root causes.
- LOOK for questions that really matter to you.

4

Coaching at Work

When my father joined the workforce, the idea of hiring a coach to help him develop his job skills would have seemed utterly absurd. His boss would have regarded such a suggestion as beyond lunacy. Why spend a penny on one-on-one coaching? You either quickly learned on the job what was expected of you or you were fired. After WW II, both public and private sector organizations began to invest substantially in staff development. The ruling model was classroom training. New employees were shipped off to training programs and given thick three-ring binders to learn. On-the-job training continued to be the dominant staff development model for small organizations as it still is today.

Today, we know much more about how adults best learn job skills than they did in the 1940s. On-the-job training and classroom programs still have their place, but computer-based training, Web casts, video-based training, teleconferences, phased learning, project simulations, and coaching have given decision makers more flexibility in how they invest in developing their people. What method is the best? It depends on the situation. No single method works for every need. However, coaching, in particular, has expanded rapidly in the past 10 years.

Coaching in the workplace, particularly executive or leadership coaching, is gaining wider acceptance. In a study released by Square Peg International Ltd. (Betchworth, Surrey, U.K.)*, the authors quote estimates that have put growth rates of executive coaching at 500% over the past five years and a similar increase over the next five years. There are three reasons behind this explosive growth: (1) **More Effective Leadership, (2) the Ability of Coaching to Deliver Results Over Time,** and (3) **Coaching Can Be Adapted to the Needs of an Individual.**

---

* www.squarepeg.com/pdf/RecentResearch/SavingExecutiveCoaching.pdf

More Effective Leadership is needed in all organizations given the ever-expanding complexities of doing business:

- **There Is an Accelerated Change in Executive Offices** around the world as aging baby boomers retire and the next generation takes over.
- **Coaching Is Unique** among the available strategies in that it can help organizations manage the risks associated with hiring and developing new leaders.

Two other factors are driving the growing popularity of coaching for developing employees further down the hierarchy.

The Ability of Coaching to Deliver Results over Time: Many of the other staff development strategies listed above are event-based. The classroom seminar or Web cast or video program is scheduled, attended, evaluated, and the process ends. They are events in time, and then they are over. Coaching can be continuous. Employees are continually involved in their own development.

Coaching Can Be Adapted to the Needs of an Individual: Have you attended a seminar that was designed to help you perform your job better, but totally missed the mark? You might have endured it by daydreaming, doodling, and watching the clock or maybe you just walked out. This happens so often because the content is not what an individual needs to learn, or the way the content is presented simply doesn't work. In a one-on-one coaching relationship, the content of each session comes from the employee. Mentally checking out is not an option.

Here's an example. A newly promoted supervisor is having trouble getting the best out of her staff team. Her boss assumes she needs a supervisory skills training program. She is sent on an expensive classroom program. Upon return, she reports that the content was very good, the facilitator interesting, and that she enjoyed meeting the other people in her class. But three months later, her manager notices that her team is still performing below par.

A few sessions with a leadership coach might have revealed that this new supervisor knows her job, has solid technical skills, and has potentially great leadership skills, but she has a problem with managing her anger. She uses cutting sarcasm to get results. She expects the worst from her people and that is exactly what she gets. So, her classroom program totally missed the mark. It may have been an enjoyable experience, but in the end it was a waste of time and money. All of which is not to say that all classroom training is worthless. Far from it.

Coaching also can turn out to be a waste of time and money. How do you go about finding the sort of coach who can best help you in your career? Look for a coach with at least three of the following factors:

- **Experience in Coaching** people like you on issues such as the ones you want to work on.
- **Certification from a Respected Coach Training Organization**, preferably one that is associated with the International Coach Federation (ICF).
- **An Intriguing Career History** that reveals an ability to take risks, cope with change, thrive on uncertainty and persist through difficult challenges to achieve success.
- **An Open, Cooperative, and Empathetic Communication Style** supported by exceptionally acute listening skills.

Remember that the person you select as a coach has got to be someone who can help you help yourself. Beware of anyone who has all the answers.

TO FIND THE RIGHT COACH, LOOK FOR

- SOMEONE WITH EXPERIENCE in coaching people like you
- CERTIFICATION and coaching experience
- A RESUME that shows tenacity, adaptability, and initiative
- GREAT LISTENING SKILLS and high empathy

5

How to Deal with Difficult People

Over the holidays, a friend (I'll call her Linda) who works in retail took me aside to vent about one of her colleagues. This individual (let's call him Stan) managed to press every imaginable button on Linda's personal control panel.

First of all, he was very talented and he knew it. Stan ranked among the top team managers in her firm. His mastery of particular product lines was unassailable. Everyone looked up to him for his expertise. Therein was part of the problem. Stan could not tolerate constructive feedback. He was always right, no matter what evidence was presented to the contrary. His insecurity drove him to dominate discussions and turn meetings into battlegrounds. He used his authority over his team to drive without mercy for perfection. Stan used the office rumor mill to sabotage people who disagreed with him. No matter how I coached her, I could not persuade Linda to come up with an effective strategy for dealing with Stan.

How can you better deal with people who drive you crazy? How can you manage these relationships more effectively so that you don't get drawn into a negative vortex that drains energy, to say nothing of precious time, from everyone involved?

Not one of the following steps is easy to put into practice. There are no quick solutions that will permanently solve these interpersonal problems. However, these steps might make life at work more bearable, even when you must work closely with a "difficult" person.

The Only Thing You Can Control Is Your Own Reaction to This Person—The person who you define as being "difficult" to work with is only difficult because of your judgment. Leave aside, for the moment, that most people have a similar reaction to this person. As in any relationship, the only thing you can really control is your own reaction. You can choose other reactions. For example, when Stan yet again derails a meeting by launching into a passionate defense of his own opinion, Linda could choose to either grind her teeth and say to herself: "Oh no, there he goes again, Mr. know-it-all!" Or she could openly address Stan: "You clearly have very strong feelings about your position in this discussion. Can you tell me more about this issue and its importance to you?" Hear him out. Let him state his case. People you define as being "difficult" to deal with are victims of our own judgment. As Stephen Covey (management consultant and author) reminded us, seek first to understand others before you demand to be understood. Stan might have some very good reasons behind his strong opinions—or not. But, unless you ask, you will never find out.

Do Not Take Their Perceived Offensiveness Personally. It's Not about You—People who press all our buttons at work have not singled us out for special punishment. What you are seeing is merely the face, or mask, they present to the world. They have learned how to use that mask long before they met you. This is their normal state, irritating though it may be. The mask you see at work is their adaptive behavior; perhaps a maladaptation due to stress or the failure to resolve other key needs, such as self-esteem. So, do not take their perceived offensiveness personally. Their behavior may have nothing to do with you or who you are. You just happen to be in the wrong place at the wrong time.

Accept the Fact That People You Define as "Difficult" Are Likely to Require High Maintenance—This means that interactions may take more time than seems justified by the circumstances at hand. Linda needs to know and plan for the fact that dealing with Stan will require more of her time and effort than with others.

Enter All Interactions with People Who You Define as Being "Difficult" Fully Grounded and Clear about Your Own Boundaries—Working with "difficult" people is really an opportunity to work on yourself. Know what is yours and what is theirs. For example, Linda needs to be clear on the boundaries she needs to have in place to cope with Stan's dominating behavior. Being clear about boundaries can allow her to give feedback in a compassionate way that might help create positive change. Linda told me about another character, this time an employee of hers, who was hypersensitive about his own perceived rights. This guy was always watching Linda to see when she made mistakes or when her behavior could be misinterpreted as being abusive or unfair. I suggested that she call him on this game and invite him to see if the two of them could get out of this negative pattern and play a more positive game.

Maladaptive Behavior Could Stem from Prior Experience in Not Having Key Needs Met. Everyone Needs Respect and Feelings of Self-Worth—Remember that Abraham Maslow's (American psychologist) hierarchy of needs (physiological, safety, love and belonging, self-esteem, self-actualization) applies to all of us, including Stan. His maladaptive behavior might come from his prior experience in not having key needs met. While none of us should attempt to play the role of amateur psychotherapist, Linda might have better luck with Stan by helping him meet some of his needs rather than getting locked into power confrontations. What might this mean at work? I asked Linda if she could recall complimenting Stan on his performance. Has she ever caught him doing something right and let him know how much she appreciated his efforts? Her answer was: "No. I thought he already had an inflated opinion of himself." He certainly displays an inflated self-opinion to the world. What about under the mask? Has anyone ever bothered to ask?

People Who Press Buttons Do So Because They Have Learned They Will Get a Predictable Reaction. Instead, Start a Different, More Positive Conversation—You can choose not to react. Instead, try to understand. Start a dialogue with questions such as these:

- "You have a lot of expertise here. Can you tell me more about your reasoning on this issue?" Build trust and self-esteem.
- "This is what I understand from what you are saying." (Then paraphrase and ask for clarification.) Build trust and understanding.
- "How do you think your reaction is affecting me/us right now?" Build insight.
- "What would you like to have happen in this situation?" Build trust.
- "All of us are going to eventually agree on a decision. What do you think would help us get there?" Encourage responsibility.
- "If you could change only two things about this discussion, what would they be?" Encourage innovation.
- "We are going to have conversations like this often. How could we conduct these discussions in a more effective way in the future?" Encourage ownership.

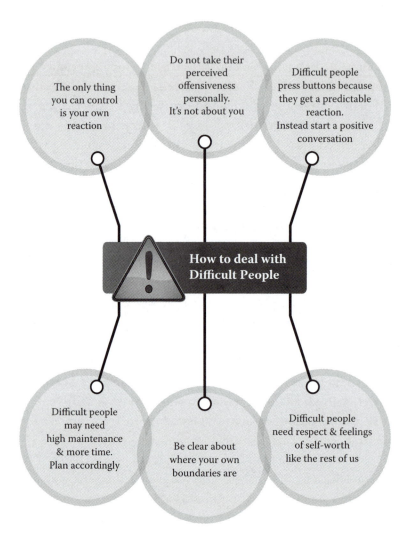

FIGURE 5.1
6 Hints for Dealing with Difficult People.

Easier said than done? Of course. But, if we are patient enough to keep our own ego needs under control when dealing with a "difficult person," we can learn a lot about them, to say nothing of learning a lot about ourselves at the same time (Figure 5.1).

HOW TO DEAL WITH DIFFICULT PEOPLE

- DON'T TAKE IT PERSONALLY. It's not about you.
- PLAN TO SPEND MORE TIME and energy with people you define as difficult.
- KNOW where you stand.
- GENUINELY BUILD their self-worth.
- DON'T LET YOUR BUTTONS get pushed. Start a more positive conversation.

6

The Gift of Business Failure: Resilience

Occasionally, in my own training and coaching business, I have gagged on the bitter taste of failure. For example: A client turns down a proposal I slaved over for days … an existing client decides on a whim to go with another supplier … I misjudge a client's situation and present a seminar that is off the mark … a large chunk of potential coaching business that I was counting on for future income inexplicably evaporates … negative feedback from one participant causes a client to cancel the rollout of a large seminar project.

When I first started out on my own, I would take these failures very personally and doubt my own competence. Could I really make it as an independent consultant running my own business? Was I in the right line of work? Why were these setbacks happening only to me? After an initial period of beating myself up, I would start analyzing the situation to find out what went wrong. Sometimes I even evolved to the point of figuring out what I would do differently next time.

It took me a few years to finally understand the true gift of failure. True, failure can teach us much about our own mistakes. Failure also can teach us humility, even self-acceptance. Failure also can help us get beyond blaming others to a place where we accept our own shortcomings. All of these are valuable gifts. However, they are dwarfed by comparison with the ultimate gift of failure, which is the ability to pick ourselves up, brush ourselves off, and get back in the game. This is the real gift of failure: *resilience*.

Unfortunately, most of us don't come into business with an innate capacity for resilience. If we are recent school graduates, we need to acknowledge that our education system has sheltered us from the essential human experience of failure. With declining enrollments and shrinking budgets, colleges and universities are reluctant to let students flunk out. When marks become unacceptably low, students are allowed to repeat courses, take academic counseling, or reduce course loads. In high school, many students are promoted to the next grade regardless of their lack of mastery of basic skills.

These trends in education mean that the younger members of our workforce can be devastated by a failure at work. Although they are talented, energetic, and far more tech savvy than their older peers, they can be severely derailed by a failure. Older workers have more life experience and tend to bounce back, simply because they have failed before and have discovered that life goes on. This resilience is particularly important in sales, where the occasional experience of failure is built into the job. Sales reps need to be able to swiftly analyze a failure, learn from their mistakes, then get back on track and make the next call. Wallowing in self-doubt costs far too much to be an option.

In business, our performance is painfully obvious to measure: the Bottom Line. Are we turning a profit and is that profit sustainable, let alone expandable? When we fail, we see the numbers and feel the pain in our bank accounts. Before we conduct a full analysis of our mistakes and decide what we would have done differently, we need to make a far more essential decision: Are we going to try again?

Failure is an emphatic teacher. There is no danger of not getting the message. When we lose money, we hurt. We doubt our own abilities, even our own self-worth. In order to try again, we need to defeat these demons by finding our own inner strength.

At the best of times in business, we experience failure as an opportunity to learn from our mistakes. Most of the time, we experience failure as a loss of opportunity, money, or a shot at success. How can we get to this place in which our faith in ourselves carries us forward with lessons learned? How can we bounce back? Are there techniques that can help us become more resilient?

While resilience cannot be learned in a seminar, here are a few ideas that might help:

1. One way to develop resilience, apart from putting in a few decades in the workforce, is to have open conversations with peers in other sectors. A frank exchange of war stories from the trenches of entrepreneurship, sales, or being an employee will inevitably yield tales of failure and recovery. Discovering how others picked themselves up and got back on track can help us become more resilient.

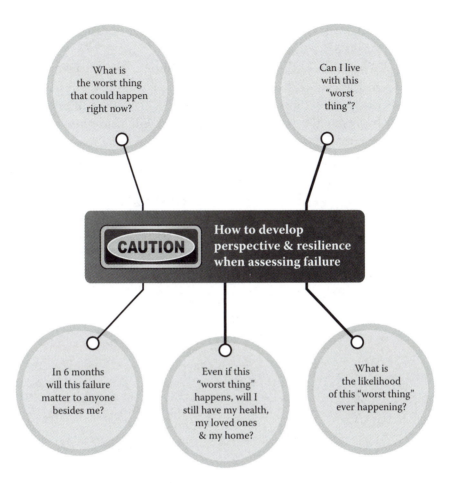

FIGURE 6.1
5 Ways to Develop Perspective and Resilience.

2. Living a balanced life also can help us ride out the stormy waters that come with failure at work. If we have caring personal relationships and supportive family ties, if we keep ourselves healthy and physically fit, and if we have interests and passions outside of work that are important to us, we stand a much better chance of recovering quickly after a failure.

3. Developing perspective, or the long view, can help us become more resilient. When we fail at work, we have to ask ourselves a few questions (Figure 6.1):

   a. What is the worst thing that could possibly happen as a result of this failure?

    b. Can I live with this worst thing, or will it destroy me (or my business)?

    c. What is the likelihood of this worst thing actually happening?

    d. At the end of the day, despite this failure, will I still have my health? Will those I love still be here? Will my life still be in tact?

    e. In six months or a year, will this failure still matter to anyone?

4. Sometimes we need to hear ourselves think in order to get through a large business or career failure. This is where professional coaching can be extremely valuable in accelerating the recovery period. A coach can help us hear what we are really saying to ourselves and help us start living from our core, that solid belief in our own self-worth.

Failure, like death and taxes, is an inevitable part of life. None of us are perfect. Don't wait for failure to hit to find out how resilient you are. Resilience is a strength you can start building today.

HOW TO MOVE ON FROM FAILURE:

- TALK TO COLLEAGUES who have been down the same road.
- LIVE A BALANCED LIFE. Constantly re-evaluate your priorities.
- DEVELOP PERSPECTIVE. Take the long view.
- FIND A COACH who can put you in touch with your own self-worth.

Section II

Leadership

7

End Continuous Conflicts at Work

Whenever I ask clients what they find most challenging about managing their own businesses or being in charge of a department in the public sector, I hear several very predictable answers, no matter the size or focus of the organization. To my continued amazement, managing workplace conflict frequently ranks among the top three issues. Consultants and facilitators, who can really help people work through these conflicts, will (unfortunately) always be busy.

A business or public sector organization is in some ways like the engine of your car. When you look under the hood, you will notice that there are many components that must work together smoothly in order for you to continue to drive. No doubt you have learned that a bit of preventative maintenance can save you a lot of trouble down the road.

Organizations, however, are more than machines. They can perform like living organisms. The health of any organism depends on fuel, rest, opportunity for growth, and a healthy immune system. If we allow internal conflict to fester like a virus unchecked, the health and even the survival of the organization can be threatened.

If you are in a leadership role, you have three choices as to how you can proactively minimize conflict:

1. You can keep pressing full speed ahead and deal with conflict as it emerges.
2. You can choose to do preventative maintenance.
3. You can take actions to boost the immune system of your organization.

Most of us would like to think that we would do the preventative maintenance and/or boost the immune system rather than wait for the inevitable trouble to hit the fan. But, often we don't have the opportunity to do the maintenance or boost the immune system. For reasons not of our own creation, we find ourselves in the middle of a mess. We acquire a business or are promoted into a position where conflict has already taken root. What can we do to stop it?

The Conflict Could Be Structural—There could be something in the design of the reporting relationships or compensation structure that generates conflict. For example, having one supervisor responsible for the performance of over 40 employees could become a structural cause of conflict. Employees might be unclear about expectations and left on their own to sort out disagreements. Either bring in another supervisor or train these employees on how to work in self-managing teams.

The Conflict Could Be Due to Inconsistent Performance Management—A performance management system with appropriate procedures and consequences might be in place, but a lax manager may choose to not abide by the rules. Such managers, eager to please everyone, inevitably wind up angering everyone. These managers need to be coached, trained, or removed.

The Conflict Could Be Rooted in Causes Outside of the Workplace—Ethnic tensions, perceived favoritism, different language abilities, and intergenerational misperceptions can all lead to conflict at work. While a work culture of high trust and respect can cut through most of these external issues, employees may benefit from knowing more about how to work with people from different generations. For example, learning how to communicate with each generation can go a long way toward innoculating against workplace conflict.

The Conflict May Be Due to a Skill Deficit in One or Two Key Individuals—Some people may be promoted to management because of their technical expertise, but they lack the people skills to do the job well. Coaching, training, and mentoring can all make a positive difference.

Do Not Play Referee—Don't put yourself in the middle of a conflict if you do not have the coaching and mediation skills to be effective. Peacemaking looks easy. It isn't. Call in help from elsewhere on your team.

Many conflicts can be minimized by proactive leadership. Take charge of the unexpected. However, if you have to react after the conflict has started, don't start at the level of personalities. First, restore the health of your organization if you can. Then you might be surprised at how quickly the conflicting personalities sort out their issues on their own.

HOW TO STOP CONFLICTS AT WORK
BEFORE THEY HURT YOUR BUSINESS

- REMOVE STRUCTURAL PROBLEMS before they become conflicts.
- DON'T TOLERATE INCONSISTENT enforcement of the rules.
- BE AWARE OF EXTERNAL CAUSES OF CONFLICT and take action to minimize them.
- MAKE SURE PEOPLE IN KEY ROLES have the communication skills to do their jobs.
- DON'T PLAY REFEREE. Work is not a football game.

8

How to Build Stronger Teams at Work

Last year I received a call from a manager of a design team at a high-tech firm. This manager complained that his people were a team in name only. In practice, they were really a collection of nine very bright solo contributors that were herded together for organizational convenience. All of them preferred to work on their own or in small cliques, collaborating across the team only when absolutely necessary. Their company, however, needed them to work closely together and speak with a unified voice to major clients. They had to be able to share ideas and develop highly integrated solutions.

At first, this manager tried fixing the situation himself. He had shipped the team off to hear an inspirational speaker. He sponsored team-building events wherein they solved problems involving blindfolds, cups of water, tennis balls, balloons, and ropes. He had put them through two expensive business simulations. He even considered taking them white water rafting for a day. But his herd of cats still refused to cooperate when they got back on the job.

Knowing that this was a very smart collection of professionals, I knew that further exposure to games, simulations, and the usual hardware of teambuilding would only deepen their cynicism and make my client's job that much harder. Even a more traditional approach involving a classroom seminar with handouts and PowerPoint® slides would just feed their negativity. "Been there, done that," I could hear them grumble.

What to do? I decided to try changing their mindsets by providing them with new information. If I could show them that there was a benefit to working together and there were things they could do to enhance their team effectiveness, perhaps they would listen to me. More importantly, maybe they would start listening to each other.

My first step was to circulate a two-page confidential questionnaire in which I asked open-ended questions about the current state of teamwork at their firm, their perceptions of the barriers and benefits to working more effectively as a team, and the actions that would have to happen in order for things to improve. I invited them to respond anonymously.

Their replies were surprising. They knew full well the benefits of working together as a team. They also knew where they were in terms of mistrust and failure to use each other as valuable resources. And, they knew what actions they had to take in order to improve as a team. Where they were totally off the rails was in their perceptions of each other. They had assumed the worst and never bothered to challenge their own assumptions.

I prepared a document containing all of their answers without attributing names, and distributed copies as the first agenda item in a one-day workshop. Being highly analytical people, they had no difficulty in identifying the major trends in the data. They knew what was wrong. They knew what they had to do to improve. Their only disagreement was over whom to blame.

Then, I assigned a self-assessment tool. This assessment tool gave them feedback on their own thinking styles. I happened to use the Gregorc Style Delineator™ developed by Dr. Anthony Gregorc (http://gregorc.com/instrume.html) because most people seem to understand their results right away. (I could have easily used other assessments, such as DISC® or True Colors®.) As they shared their results from the Style Delineator, they quickly realized why they had so much difficulty in trusting each other; they really saw the world in very different ways.

Some of them were highly structured in their perceptions. Everything had to be done in an ordered, logical way. Others were very creative and random in how they worked. These people detested rules, procedures, and bureaucracy. Others were very patient and analytical. And there was one individual who lived in a constant "just-in-time" state of hyperalertness. She had no patience with considering options, deliberating, or analyzing; she was only interested in getting results, pronto. All of these thinking styles are valuable, but they have to understand each other.

Next I asked the team to divide into the four main thinking styles and asked them to describe how they would make an imaginary presentation to someone of an opposite style. The very logical, ordered people had to present to the creative types and vice versa. I compelled them to walk in someone else's shoes for a few minutes.

As soon as this team understood that they perceived and organized all information in very different ways and, as a consequence, their disagreements were not resolvable, they began to laugh at themselves.

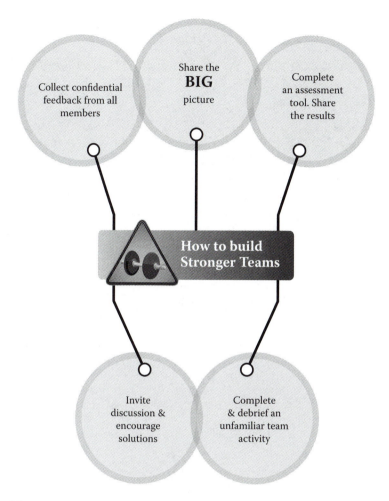

FIGURE 8.1
5 Actions for Building Stronger Teams.

Then I invited them to complete a complex team activity in which they had lots of opportunity to practice their new understandings of their different thinking styles. Now they were ready to learn as never before. We finished the day by making action plans about how to work better together.

The combined effect of the survey, the assessment tool, and the team activity was to give this team a greater sense of tolerance and acceptance. They taught me that teams, with just a bit of help, often know what is best for themselves, and, as managers, we need to know when to get out of the way and trust the team process (Figure 8.1).

HOW TO BUILD STRONGER TEAMS

- COLLECT CONFIDENTIAL FEEDBACK on all members as a collective unit.
- SHARE the big picture.
- COMPLETE AN ASSESSMENT TOOL. Share the results.
- COMPLETE AND DEBRIEF an unfamiliar team activity.
- INVITE DISCUSSION and solutions.

9

Catch Your People Doing Something Right

In organizations across the country, management continues to exhort their employees to identify and overcome weaknesses in order to achieve optimum performance. The Holy Grail of career success, we are told, is to be found in the heroic language of struggle, overcoming challenges, breaking through barriers, and working ceaselessly to develop a broad portfolio of competencies.

Peel away the jargon and we find that most workers are still being told that, if only they could get better at what they are not doing so well right now, then someday they would really get ahead. Yet, experience tells them something different. Many find that they cannot really improve their weaker competencies beyond a rather modest midpoint. Try as they might, they can't seem to crack the magic code.

According to Marcus Buckingham, we are trying to solve the wrong problem. Buckingham, along with Donald Clifton in their book, *Now, Discover Your Strengths* (Free Press, 2001), makes the truly radical proposal that when it comes to developing people, management gurus have been getting it all wrong for decades. The path to High Performance Heaven, these gurus tell us, lies, in part, in identifying the competency profile of a mythical set of ideal performers, assessing employees to find out where they fall short of this ideal, then challenging them to overcome their weaknesses as they strive for the Promised Land of Perfection.

This competency model of leader and staff development has a long history. The military establishments of the United Kingdom, the United States, and other NATO countries adopted this model for their leadership development programs during the 1950s and 1960s. The leading military thinkers of the day believed that if they could identify the perfect competency profile of the ideal officer, then train all their officer candidates in these competencies and measure their progress against this ideal, that they would generate a cadre of top-tier leaders. But the model didn't work. The military found that most people could not improve their performance on all the ideal leadership competencies, particularly their weaknesses, to a sufficient level. So, they abandoned this competency model.

Unfortunately, no one in the business world noticed. Thus, today we still hear much inspirational chest-thumping about leadership competencies from business and government leaders. Too bad. It was a grand theory, but not so useful in the real world.

Buckingham and Clifton suggest that if all of us became experts at finding, developing, applying, and marketing our strengths, then we would be taking a crucial first step toward building strengths-based organizations. What specific actions could leaders take to convert their organizations from job descriptions to strength profiles so that their people could do what they do best of all and get paid for it? (Figure 9.1)

BUILDING A STRENGTHS-BASED ORGANIZATION: LEADERSHIP BEST PRACTICES

Talk to Your People and Find Out What Their Strengths Are—Find out what they bring to the table. Encourage them to share the talents and skills they use outside of work as well as on the job. You will never know what gold is buried somewhere in your team if you don't ever take the time to do some serious prospecting.

Champion the Use of Assessment Tools that help people identify their talents. A great place to start is the online assessment referred to by Buckingham and Clifton in *Now, Discover Your Strengths*. Online at: http://www.strengthstest.com/now_discover_your_strengths.php

Support the Development of People Who Want to Convert Their Talents into Strengths—That means supporting their efforts to take part in mentorships, cross-functional assignments, coaching, training, self-development, and professional development events, such as conferences.

The converse of this action is to **Identify and Confront the Drifters**. These are people who are really only going through the motions at work. Doing the minimum. You and they cannot afford to bring less than 100 percent of their commitment and talent to work every day. Try assigning

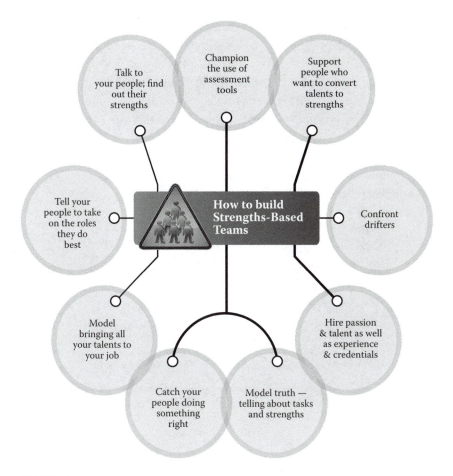

FIGURE 9.1
9 Steps to Building a Strengths-Based Team.

them to other tasks. If a change doesn't work, you may have to let them go. As difficult as these decisions are, your forthright action may well kick-start these individuals into an ultimately rewarding confrontation with existential career issues, such as: "What do I want to do with my career?" and "What do I do best?" Down the road they may even thank you.

Hire Passion and Talent as well as track record and credentials. Look for demonstrated passions; ask for evidence of prior activities or accomplishments using the talents being claimed by your applicants.

Encourage and Model Truth-Telling about the Fit between Tasks and Strengths—Reflect on your own career history; find anecdotes to support the value of going for what you really, really want to do and what you are really good at.

Catch Your People Doing Something Right—Reward exemplary demonstrations of talent. Encourage these exemplars to develop their talents into strengths.

Encourage and Model, Bringing All of Your Energies and Talents to Your Job, not just the ones that fit into a formal job description.

Instead of designing jobs around functions, **Ask Your People to Take on the Roles and Tasks They Do Best**. This may sound like risky advice, but the more that your people use their natural talents at work, the more your people will take genuine ownership for the business success of your organization.

HOW TO BUILD A TEAM BASED ON STRENGTHS

- ENCOURAGE YOUR PEOPLE TO GET BETTER at what they are already good at.
- STOP TRYING TO FIX individual weaknesses.
- HIRE PASSION AND TALENT in addition to credentials and experience.
- CATCH YOUR PEOPLE doing something right.
- REWARD your stars.
- BUILD JOBS around what people do best.

10

How to Convert Enemies to Allies

Not long ago a director of a midsized consumer goods retailer asked me to help her resolve a conflict between two of her managers, Sarah and Carl.

Sarah

Sarah was focused, driven, and had a reputation for getting what she wanted, no matter what. Her most important project, the project upon which her bonus depended, was in jeopardy. Despite her sincere efforts at negotiating in good faith, she could not get Carl, the manager of Information Technology (IT) for her region, to support her. Her project required a significant level of teamwork. Everyone in the region would have to cooperate if her marketing project was to succeed. Most of all, she needed Carl's 100-percent support.

Sarah complained, "Look, I have done everything possible. I have been patient. I spent hours in meetings. Nothing has worked. Carl just cannot and will not listen. The man is a train wreck and he is destroying this company, to say nothing of my project. I can't go over his head to the management team. They would rightly say, 'Sort it out yourself.' What else can I do? He is behaving like a complete idiot for no reason at all.

"Carl says he doesn't have the headcount to take on my project. Despite my detailed description of how the project would showcase his department across the region, despite my assurances that other resources from other regional IT departments could be brought in for the duration of the project, he continues to stonewall my ideas. I've been reasonable, I've been fair, and I have even suggested bringing in outside contractors to help him. Instead of seeing this suggestion as a value-add (he gets the credit for the work of others and their costs don't come out of his budget), he gets all defensive and huffy. He is more territorial than a junkyard dog! I am so finished with him. Done! I've had it dealing with that clown. But, now what will I do?"

Sarah knew she was on the hook for delivering her project on time. Her deliverables included a complete marketing plan that could be rolled out across the region and which built on her firm's growing marketing presence on social media sites, such as Facebook. She desperately needed Carl's help and at that moment such help was definitely not on offer.

What could Sarah do?

- First off, she needed to focus on the problem (finding the IT resources to support her marketing project) not on the person with whom she was experiencing so much frustration. Her problem was not Carl.
- Sarah needed to find other influencers who could have talked to Carl.
- Next, she needed to stay riveted on her desired outcome: a successful marketing project. As the success of that project was what she really wanted, what should she have been doing to increase her chances of achieving her goal?
- She needed to better access her internal support network. Who else might have been able to help her access the technical expertise she needed to achieve her results?
- She needed to reframe her project in terms of benefits to her peers. Instead of rigorously pursuing her own objectives to the detriment of everyone around her, she needed to reposition her project as a team effort that would benefit everyone. She needed to ask herself how her project could have helped Carl achieve his objectives, and she needed to look for ways Carl's participation in this project could have enhanced his profile with the management team.
- She desperately needed to get out of her "my project" frame of mind and embrace an "our project" mental model.
- Sarah needed to improve her internal selling skills. She needed to create alliances, look for common cause, uncover shared interests, and, most of all, embrace the language of benefits. She needed to be able to clearly answer everyone's WIIFM question (what's-in-it-for-me).

Hell might freeze over before Carl would have taken time out of his hectic schedule to accept more work as a favor to Sarah. However, he might have become involved in an exciting project that everyone was talking about and which could have significantly benefited his own career.

Carl

Of course, Carl saw the problem quite differently; he could not see the benefits of Sarah's project. Carl was 10 years older that Sarah. He had a solid reputation as a fair-minded pragmatist who treated people decently, but never lost sight of what was best for the business. He was an excellent IT professional, but had no time for unproven technologies or "techie dreamers," as he called them. Moreover, he was highly exasperated by Sarah's assertive approach. The fact that Carl had recently been through a messy divorce and was not fully comfortable working with assertive female peers did not help. The more she pushed, the deeper he dug in his heels. Given Carl's anger, what could Sarah have done to change this game? And what did Carl risk by not supporting Sarah?

When I met with Carl, he was clearly in a mood to rant.

"Look, I wasn't born yesterday. Sarah doesn't set my performance objectives. My director does that. One boss is enough. I will not let that woman push me around. In my department, we have our objectives and more than enough projects on the go. I am not going to risk overreach or failure on any of my deliverables just to satisfy her ego. Her project, this social media marketing caper, is nothing more than an attention grabber. Anyone who has been in this business for half as long as I have can see that this is just power politics, pure and simple. She needs a big win, something really sexy, to raise her profile with the management team so she can jump queue for the senior manager's job. Never mind that her project makes no business sense or that it wastes a lot of time and money en route. Never mind that her project could thoroughly trash my chances of meeting my own targets this year. What does she care? If she is so committed, let her find the budget to hire her own consultants and make it happen. Let her spend her own bucks. Then she has the nerve to fill up my schedule with meetings to try and humiliate me in front of my peers and goad me into getting on her side. Unless my boss tells me to help her and adjusts my current workload to free up enough capacity, to say nothing of budget, she's on her own. I have more than enough on my plate. But, she can't hear that. To heck with her!"

- While Sarah could not stop being who she was, she needed to avoid pushing Carl's buttons unnecessarily.

- Carl needed to get more distance between his own reactions to Sarah's style and the reality of what she was asking of him. He needed to get more control over his own buttons.
- Carl needed to get control of his own internal monologue. Sarah was not the devil. She was just another leader like himself who was passionate about getting results. His trashing of Sarah, even if he kept it only to himself, was not going to help him now or in the future. Believing the worst about people leads to poor teamwork, and Carl needed to be able to play on the same team as Sarah.
- Carl needed to consider the possible upsides of helping Sarah succeed. He needed to embrace the potential of this exciting new project for enhancing his own career while costing him a minimal amount of time and effort.
- Carl and Sarah both needed coaching to help them clear the air.

The truth does not lie in taking sides. Both Sarah and Carl needed to own their respective contributions to this conflict. Resolving conflicts between peers requires patience, attentive listening, and, above all else, a deep faith in people to succeed. A solution that people create by themselves is much more lasting than an imposed solution from an authority figure.

HOW TO CONVERT ENEMIES TO ALLIES

- FOCUS ON THE ISSUES, not the people.
- ALWAYS KEEP THE DESIRED OUTCOME in mind.
- SELL ENEMIES ON THE BENEFITS of your proposal to everyone.
- HELP EVERYONE get the results they want.

11

What Really Motivates Employees? A New Approach to Motivation

For anyone else who manages staff, getting the best out of your people can be a continual struggle. How can you reward good performance in a way that will guarantee continued improvement? How can you ensure that poor performance is turned around? Even during a recession, fear of losing one's job, at best, only leads to compliance with minimum performance standards. Sullen compliance is no substitute for enthusiasm and excellence. Fear is not a very good motivator of discretionary effort.

When I joined the workforce, my performance was managed very simply: I received recognition or even bonuses when I did well (those were the carrots) and I was chastised or threatened with being fired if I messed up (those were the sticks). "Carrot-and-stick" motivation has been with us for centuries, and for good reasons. Up to a point, it works. Soon after the Industrial Revolution transformed nineteenth-century workplaces, managers discovered that workers doing routine physical tasks, such as making bricks, worked harder when offered rewards for good performance, and poor performance could be improved if dire consequences were threatened or enforced.

We, however, are no longer in the nineteenth century. Very few of us earn our pay by performing routine physical tasks, day after day. We are in the third wave of the Information Revolution. Yet, we still use the same outdated carrot-and-stick model of motivation. Many (but not all) workers these days have levels of problem solving, innovating, continuous learning, and decision making in their jobs that their great-grandparents could have never even imagined.

What happens when we apply an industrial-age model of motivation to information-age performance challenges? It just doesn't work. Or, it doesn't work as well as it used to. If you offer me a substantial cash incentive to make more bricks per hour, I will rise to the challenge, at least for a while, or until I can find another employer who will offer me a better deal. However, if you offer me the same cash incentive to solve a thorny operations problem or create more efficient ways of making products or delivering services, the carrot-and-stick model breaks down. I'll stay late working at the problem because I want the reward, but I won't necessarily come any closer to producing the result of which you had hoped.

Daniel Pink, author of *Drive: The Surprising Truth About What Motivates Us* (Riverhead Books, 2009) (online at: http://www.danpink.com/drive), suggests that to unleash the full capacity of today's employees, particularly those working on creative or complex tasks, managers need to respond to three major employee aspirations:

1. **Autonomy:** They want to be in charge of their own work.
2. **Mastery:** The opportunity to become as good as they can be at their jobs.
3. **Purpose:** They want to work on something that matters to them.

Autonomy implies having significant control over how you work, what you do, when you work, and with whom you work. Mastery, for Pink, means having the chance to learn, to improve at tasks, to excel, and to receive frequent performance feedback. Purpose means doing worthwhile work. Purposeful work touches something greater than the individual; this work matters to others. Most of all, such work really matters to the individual employee.

Some employers who embrace Pink's ideas are making bold changes in how their employees get their work done. In creative fields, such as advertising, design, technology, and media, some companies have even thrown out the concept of regular hours. Managers say, "For the most part, I don't care when you work. I don't even care where you work, as long as I can get in touch with you right away. But, I expect you to get your work done on time. How that happens is up to you."

While the need for autonomy is perhaps the easiest of these aspirations to address, the need for purpose is far more difficult to meet. Pink describes one employer who tells his staff that as long as they are getting their work done, they can spend one day every week working on whatever they want to as long as this work is not related to their immediate jobs. The participating employees must report back the next day on the results of their freewheeling efforts. The company has experienced a windfall of new product ideas, fixes to long-standing customer service issues, and innovative research projects that otherwise would have never emerged. Some of these employers also encourage their people to get involved in meaningful causes and campaigns during work hours.

What about money as a motivator? All of the above ideas are based on the assumption that employees are paid enough. Once we pay people adequately, depending on their position and tasks, paying them more doesn't necessarily lead to an equivalent improvement in performance. Of course, money still matters, but beyond our basic needs, it may not matter as much as we think.

While it may seem relatively easy to redesign a motivation strategy for a group of software designers around the three aspirations of autonomy, mastery, and purpose, what about a group of janitors? How can they be motivated by these three aspirations? Pink describes an application of his ideas to a team of hospital janitors. These workers were always told to clean a number of rooms in a set order within a fixed number of hours. Their jobs were changed to encourage them to interact more with patients and to be more of a social presence on the wards where they worked. The result was that these janitors felt a deeper connection with their jobs and the importance of their work in maintaining a healing environment.

Daniel Pink's ideas make sense to us because they ring true in our own experience. We work best when we are treated as adults worthy of respect, who can be counted on to deliver results. We work best when the rewards are intrinsic; we do the work because we find the work itself to be fulfilling. That doesn't mean that we don't want to be paid well.

Managers who understand this sense of intrinsic motivation compensate their people in a way that removes money as an issue so they can focus on their jobs. These managers create workplaces where people want to do their best. This means giving them autonomy over their time, tasks, and teams, encouraging them to really master their jobs and giving their work a sense of greater purpose. These managers don't do this just because it feels good. They do it because it leads to better business results.

In the end, what seems to work best is getting the combination of extrinsic rewards (pay and benefits) and intrinsic motivations (autonomy, mastery, and purpose) right. The best places to work pay well *and* treat their employees like the competent, trustworthy adults they are. All business owners and managers like to say, "Our people are our greatest asset." To get the very best out of our people, we need to act like we really believe those words (Figure 11.1).

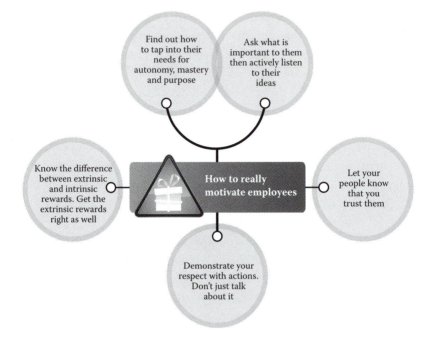

FIGURE 11.1
5 Ways to Really Motivate Employees.

HOW TO REALLY MOTIVATE EMPLOYEES

- FIND OUT HOW TO TAP INTO THEIR NEEDS for autonomy, mastery, and purpose.
- ASK WHAT IS IMPORTANT to them, then actively listen to their ideas.
- LET YOUR PEOPLE KNOW that you trust them.
- DEMONSTRATE YOUR RESPECT with actions. Don't just talk about it.
- KNOW THE DIFFERENCE BETWEEN EXTRINSIC REWARDS AND INTRINSIC REWARDS. Get the extrinsic rewards right as well.

12

Recession Survival Strategies: Courage and Entrepreneurship

Whether you are a business owner, an employee, or a solo entrepreneur, you will need to be at your best to survive this recession. Victim behavior, blaming others, the bunker mentality, and obsessive cynicism are all strategies for failure. Regardless of your own financial predicaments, the best person to get things moving is patiently waiting for you in the mirror.

There is more than enough gloomy news to discourage everyone. We all know the script: bank failures, stock market crashes, credit crunches, job losses, and an unprecedented collapse of consumer confidence. Not that you should hide from bad news. Instead, you need to confront it and try to understand it. You, however, should not wallow in this bad news. You need to move beyond passive news consumption and take action. Now perhaps more than any time in the past, you need to arouse your own sense of entrepreneurship.

The way of the entrepreneur is not easy. At a time when circumstances cry out for bold action, employees often shrink into a self-preserving shell of hypercautiousness. Business owners who need their staff to come up with new ways to compete and succeed instead are confronted with apathy, fear, and hopelessness. How can you turn the corner?

What does it take to work like an entrepreneur?

Know Yourself: First you have to know your own strengths, be willing to believe in your own dreams, have faith in your ability to create the results you want, know what is most important in your life, and know what you want to achieve.

Know Your Brand: What is your brand? Where do you stand? If you had 20 seconds in an elevator to describe your brand, what would you say? What stories of your life and accomplishments will you want people to talk about at your funeral?

Drive: Whether you are a leader or an employee, successful entrepreneurship requires great energy and drive. You have to get in touch with what you really want and why it is so important to you. You need to be able to paint a picture of your own success in your mind, then be ready to put all of your energy behind the goal of converting your dream into a reality. You have to be hungry for success.

Values: You have to know what your core values are and what is most important to you. Why? Because these core values will leave their mark on everything you do. These values have to be aligned with who you really are and what you really want to create. And others will have to find your values compelling.

People: Entrepreneurs need to attract and convince others. They have to understand how to motivate employees, attract investors, satisfy clients, sustain mutual trust, and treat everyone with respect. Other people must want to get on your bus and want to travel with you. So, you have to know how to listen and how to motivate others.

Honesty: Successful entrepreneurs start from where they really are, not from where they would like to be. They are brutally honest with themselves and invite honest feedback from partners, employees, and clients. An entrepreneur with a bad idea wants to know what is wrong.

Initiative: Great achievements in business do not happen without a strong willingness to get started and to take action, even if the action is not precisely the best action possible. We have all read the stories of legendary entrepreneurs who started as taxi drivers, bought their own cab, borrowed more money, bought a share in a small business, then leveraged their success into a significant business empire. Are those entrepreneurs really all that different from the rest of us?

Courage: Can courage be learned, or is it something with which we are born? Dr. Merom Klein of The Courage Institute (Baltimore, Maryland) (http://www.courageinstitute.org/) maintains that courage is a choice we can all make about how we deal with the adversity around us. All of us face fears, but we need to learn what we should and should not be afraid of, then how we can face those fears constructively. Successful entrepreneurs are very skilled at separating useless fears from those they need to confront and overcome. Fear of bankruptcy is real; all entrepreneurs need to confront this fear and deal with it. Fear of market rejection in launching a new business can, on the other hand, keep a potentially successful entrepreneur frozen in a stultifying career that leads nowhere.

Taking Risks: Along with confronting fear, entrepreneurs need to know how to take the big risk—the risk that will either launch them on the road to success or cause them to fail and have to start all over again. Successful risk takers, whether in business or any other field, need to know how to manage risks and take control of as many of the variables as they can. Successful entrepreneurs who take risks are sometimes very cautious individuals. They leave as little as possible to chance. They conduct detailed risk assessments, weigh all their options, do their research, and design backup systems in case of trouble. But, in the end, they stand up and move forward. They take the risk.

If you are a business owner or a manager, you can help your people become more entrepreneurial. Entrepreneurship may or may not be teachable, but it can be encouraged (Figure 12.1).

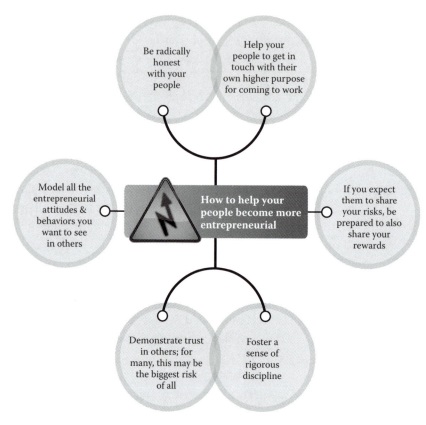

FIGURE 12.1
6 Actions for Building a More Entrepreneurial Staff.

HOW TO HELP YOUR PEOPLE BECOME MORE ENTREPRENEURIAL

- MODEL ALL THE ENTREPRENEURIAL ATTITUDES and behaviors you want to see in your people.
- BE RADICALLY HONEST with them.
- HELP YOUR PEOPLE TO GET IN TOUCH with their own higher purposes for coming to work.
- IF YOU EXPECT THEM TO SHARE YOUR RISKS, be prepared to also share your rewards.
- FOSTER a sense of rigorous discipline.
- DEMONSTRATE TRUST in others; for many, this is the biggest risk of all.

13

What Frontline Managers Need to Know about Delegating Work

When I worked for a bank years ago, I had two memorable managers. Gail was a wonderful manager. She always seemed to know how to get me to do my best work. I didn't know how she did this and, to be fair, she didn't know either. Our working relationship was trusting and respectful. I enjoyed my job and felt as if my work mattered.

The less I say about working for Harry, the better. He used a passive-aggressive style of leadership and complained about me behind my back. What he said was not what he meant. I would ask for feedback on my projects, get none, and then be told I was going in the wrong direction. Going to work each morning for Harry was like serving prison time.

With the wisdom of hindsight, I can see now that there was much more going on between these two managers and myself than could be explained simply by personality. The differences between these two had to do with how they delegated tasks. Gail expected the best out of me, and then set me free to deliver just that. Harry expected the worst and that was often what he got.

Now I can see that Gail intuitively understood three things about everyone she managed:

1. **Capability:** The degree to which her people had the skills and knowledge to do what she wanted to assign to them.
2. **Desire:** The degree to which her people wanted to do the work she needed to assign to them.
3. **Task Ownership:** The amount of ownership for an assignment that an individual was prepared to willingly accept.

Three examples include:

#1: Retail Supervisor

Let's suppose you are a supervisor in a retail outlet and you need to assign a complex inventory task to a new employee. This new employee really wants to succeed, but he doesn't fully understand the inventory management software at your store.

What is his level of capability, desire, and task ownership? How would you assign this task to this person?

- **Capability:** Low, because he is new. This means you will have to be very deliberate as you teach him how to use the inventory system.
- **Desire:** High. In terms of managing his desire, you may have to boost his confidence a bit so that he does not feel intimidated by the task.
- **Task Ownership:** Low. You will have to monitor his progress closely. Be directive and supportive.

2: Manufacturing Supervisor

Suppose you are a new supervisor about to assign a task to a very experienced machine operator in a manufacturing plant. This operator knows his job and he can do good work, but is easily bored. You need to assign him a custom-design task for an important client. You know he has done similar tasks before and enjoys the challenge of doing something new.

What are your operator's levels of capability, desire, and task ownership? How would you assign this task to this person?

- **Capability:** High. He knows what he is doing. Instead of being very directive and boosting your operator's confidence with a little pep talk, you should treat this operator as a very responsible peer who has the capability to get this job done right.
- **Desire:** Moderate. It wouldn't hurt to tactfully remind him that this is an important client.
- **Task Ownership:** Moderate, not because he doubts he can do the job, but he may not get fully engaged with the task until he gets into it. You should transfer responsibility for this job to him, then ask him to get back to you on his progress later. He can do the job and get it right. Empower him. Show that you trust him.

3: Hospitality Manager

Now suppose you are a new manager in a hotel kitchen and you must ask an experienced employee to clean a large appliance. This employee knows how to do the job, but does not like cleaning.

What are this employee's levels of capability, desire, and task ownership? How would you assign this task to this employee?

- **Capability:** High. She knows how to clean the appliance, so don't bother to explain how to conduct the cleaning procedure.
- **Desire:** Low. Summarize the importance of cleaning in terms of food safety and government inspections. She knows this, but a reminder is a good benchmark. Then acknowledge that cleaning tasks are equitably distributed among all kitchen staff and that today is her turn. Describe any consequences that might affect the kitchen's performance, such as the ability to produce all the food for the banquet scheduled for the next day.
- **Task Ownership:** Low. Tell her that you will be close by, working in the kitchen and can replace any cleaning supplies as she runs out if she asks you. Stay in control without hovering over her. Back off as soon as you see progress.

The next time you delegate a task, be sure to analyze the level of capability, desire, and task ownership of the person who is going to do the work (Figure 13.1). The outcome might be a better job done and a better workplace for everyone.

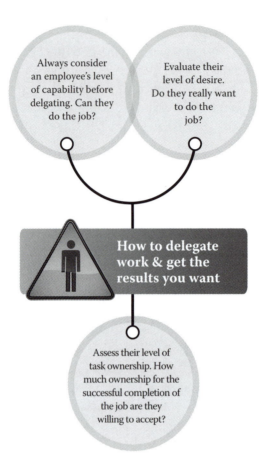

FIGURE 13.1
3 Steps for Delegating Work.

**HOW TO DELEGATE WORK AND GET
THE RESULTS YOU WANT**

- ALWAYS CONSIDER AN EMPLOYEE'S LEVEL OF
  CAPABILITY before delegating. Can they do the job?
- EVALUATE THEIR LEVEL OF DESIRE. Do they really want
  to do the job?
- ASSESS THEIR LEVEL OF TASK OWNERSHIP. How much
  ownership for the successful completion of the job are they willing to accept?

14

The Science and Art of Great Interviewing

As the so-called "green shoots" of economic recovery continue to pop up, many businesses, both large and small, may need to hire more staff to cope with increased customer demand, or so we all hope. This could mean that managers might be doing a lot of interviewing.

Conducting effective interviews—interviews that result in finding the right people for the right price—is an incredibly important action for driving business success. A lot of benefit—and risk—is riding on those few minutes an interviewer spends with each candidate. You could hire a dedicated genius that will make you extremely successful. Or you could hire a chameleon that, once onboard, turns into an incredible liability, costing you money, time, and customers, never mind all the unnecessary stress. Great interviewers consistently hire the right people. Conversely, they usually don't hire the wrong people. How do these great interviewers get it right?

They know that effective interviewing is part art and part science.

EFFECTIVE INTERVIEWERS ARE GREAT AT ASSESSING THE 3 CS: CAPABILITY, COMMITMENT, AND CHEMISTRY

While some of us have a knack for building rapport, encouraging job candidates to be open and asking great questions, all of us can learn how to be better interviewers. All it takes is practice, feedback, and checking for the 3 Cs.

- **Capability** *(Behavior: Has the candidate done tasks like this in the past?)* This is the degree to which the candidate has successfully performed the skills, demonstrated the behaviors, and obtained the results required for the position in the past. Past behavior is a fairly reliable—although not perfect—predictor of future performance.
- **Commitment** *(Motivation: Has the candidate demonstrated the appropriate drive and the will to do similar tasks in the past?)* This is the level of motivation the candidate has for the advertised position and all that is entailed in performing effectively in that position.
- **Chemistry** *(Attitude: Will the candidate fit in with your organization's culture? How compatible will he/she be with the other people on your team?)* This is the quality of fit between the candidate, the culture, the way current employees work together (including abiding by decisions), making decisions, adhering to standard procedures, negotiating, communicating, solving problems, and resolving conflicts.

Before the interview, you must prepare questions that will allow you to gather evidence for making a hiring decision based on these 3 Cs. Capabilities are fairly straightforward to evaluate. Commitment and Chemistry can sometimes be overlooked, in part because they are more difficult to assess. Some experts maintain that the bulk of poor hires do not work out because of inappropriate Chemistry (attitudes).

Use a Consistent Structure for All Interviews

Follow a consistent structure in planning for, conducting, and evaluating your interviews. A consistent structure will ensure that

- you get a **complete and balanced picture** of the skills, behaviors, motivations, and attitudes a candidate brings to the position;
- you will be able to **evaluate all candidates using the same criteria; and**
- you will be able to conduct your interview in a professional manner and **abide by all legal requirements.**

Prepare the Right Questions in Advance

- Create questions that will lead to the candidate giving you specific evidence as to his or her on-the-job Capability, Commitment, and Chemistry. Try to come up with a minimum of four questions for each of these categories.

- Test your questions with a peer or trusted employee who has performed well in the position for which you are conducting the interview. Make sure your questions ask for specific information. Do they make sense? Will these questions lead to the information you need?

In general, avoid questions that will yield only general information. *Remember that you are gathering evidence for a hiring decision.* Your questions must reflect this need to surface specific information on which to base your decision.

During the Interview

Build Rapport and make the candidate feel at ease. A job interview can be a very stressful experience. Some candidates will need a lot more rapport building than others.

Describe the Structured Interview Process. Tell the candidate that you will be following a structured process in order to find out more about what he or she can do and how he or she might fit into your company.

Explain That There Will Be an Opportunity to Ask Questions at the end of your questions and acknowledge that a job interview is a two-way process.

Ask Your Prepared Questions. You should use a printed question sheet with room for answers below each question and a scoring system for ranking each response.

Keep the Interview on Track. Control the Agenda and do not hesitate to draw an overly talkative candidate back on track.

Be Aware of Your Own Bias. You may, sometimes for unconscious reasons, really like a particular candidate or really dislike another.

Do Not Overlook the Need to Check for Fit into Your Organization's Culture

Ask for Both Sides of an Issue. For example:

"Please give me a specific example of when you used good leadership skills to resolve a team conflict." And … "Can you give me an example of a time when you could have used better leadership skills in order to resolve a team conflict. What happened and what did you learn from this?"

The Candidate Should Do Most of the Talking, around 80 Percent. Give them room to talk. **Don't Be Afraid of Silence**. Give candidates time to think.

You will interview highly honest people, highly unscrupulous people, assertive individuals with a gift for self-promotion, people who are very reluctant to draw attention to their accomplishments, raving egotists, extremely shy people, and everyone in between. Use a structured interview process to protect yourself against the bias that may come from your own reactions to this huge range of humanity (Figure 14.1). For more information, see

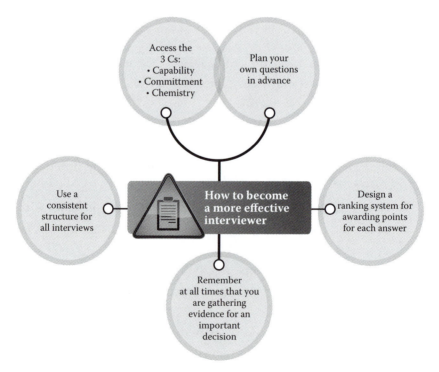

FIGURE 14.1
5 Ways to Become a More Effective Interviewer.

the excellent DVD, "The Three-Dimensional Interview," from VisionPoint Productions. Online at: http://visionpoint.com/training-solutions/the-three-dimensional-interview-skills-training-for-hiring-managers/

HOW TO BECOME A MORE EFFECTIVE INTERVIEWER

- USE A CONSISTENT STRUCTURE for all interviews.
- ASSESS THE 3 Cs: Capability, Commitment, and Chemistry.
- PLAN YOUR QUESTIONS in advance.
- DESIGN A RANKING SYSTEM for awarding points for each answer.
- REMEMBER AT ALL TIMES THAT YOU ARE GATHERING EVIDENCE for an important decision.

Interviewing as a Science:	Interviewing as an Art:
Great interviewers are highly disciplined and they always: **Prepare** for an interview well in advance and **Follow a Set Structure** during the interview. **Keep Track of Their Observations** and use a structured evaluation process.	Great interviewers are excellent listeners and can read people very well. Above all, they are: **Fully Present** and very attentive during the interview. Able to **Trust Their Intuition** as to when to probe and when to ask open or closed questions.
Great interviewers stay focused on the results they want to obtain: **Solid Evidence to Back up Their Assessment of the Interviewee.**	Great interviewers can: • **Put People at Ease.** • **Show Respect.** • Keep the conversation on **a Business Footing.** • **Always Be Fully Aware** of their biases.

Interviewing as a Science:	**Interviewing as an Art:**
Great interviewers use a **Structured Process** for assessing each interviewee. Great interviewing depends on a **Methodical, Precise Approach** that ultimately is fair to all the job candidates being interviewed.	Great interviewers are also **Empathetic** in that they acknowledge that a job interview can be a highly stressful event and that some candidates may need time to think before answering.
Scientists in a lab must subject all of their specimens to the same testing procedures in order to conduct a balanced assessment. So, it is with interviewers and the candidates they interview.	At all times, they **Display Good Judgment.** And, they are very careful to avoid any questions or discussions that could reflect poorly on their employer. **They Abide by the Law.**

Capabilities	
"Can you do this sort of work?"	You will get a yes or no answer, or possibly something rather sketchy in between. You won't get much useful evidence.
"Please give me two examples of work you have done similar to what I just described." "Where did you do this work and how did you know you were successful?"	The candidate cannot hide behind generalities here. He or she will have to be specific. You will get much more useful information.

Commitment	
"Suppose you were given a rush assignment late on a Friday afternoon before a long weekend. How would you handle this situation?"	The candidate can always give you the answer you want to hear. Even if they answer honestly, you won't get much useful evidence. You will get an opinion. You want facts.
"Give me a specific example of a time when you had to change your schedule or work overtime in order to accommodate an emergency request from your supervisor."	As above, the candidate has nowhere to hide. True, they can always make up situations. In that case, you can ask specific questions to probe deeper.

Chemistry	
"How well do you get along with supervisors and managers at work?"	Again, the candidate may give you an answer that he or she thinks you want to hear. By itself, this is not a useful question.
"Describe your relationship with your former supervisor at your last job. Please give me a specific example of something that caused a problem or issue that the two of you had to resolve. What happened?"	The candidate has to come up with a specific example of something that happened in the last job. You will get more useful evidence.

15

Managing during Tough Times

To say that these are very tough times to be a manager, no matter what the business sector or size of organization, is more than just an acknowledgement of the obvious. The game has changed. What worked before may not work anymore.

First, there are the client issues: cancelled orders, plans for expansion put on hold, decision-making processes that seem forever caught in limbo, pleas for special deals on pricing, protracted negotiations, dropping business volumes, clients facing business failure or credit issues, and a pervasive tendency to be overly cautious about spending and anxious about what the next several quarters may bring.

Then, there are the challenges of managing worried employees. Managers must treat their staff with utmost care during this time of constant upheaval in the economy. Stress levels climb daily as wave after wave of bad news is delivered on the morning news. Older employees who were considering retirement now are making other plans as they watch their savings shrink. Younger employees, eager for promotion, are caught in an increasingly tight bottleneck by these reluctant retirees. As business sectors, such as manufacturing, tourism, and forestry, continue to get hammered by layoffs, employees begin counting the job losses among their friends and families. Adult children unable to find good-paying jobs are moving back home with mom and dad. Young graduates are unable to break out of minimum wage positions while their school loans hang overhead, mostly unpaid. House prices have fallen and sellers must be much more patient than they were a few years ago.

What strategies should managers and business owners adopt in order to get through the next few quarters and possibly the next few years of diminished expectations? Here are two ideas.

SCCIO ANALYSIS

Do a Situation–Causes–Consequences–Implications–Opportunities (SCCIO) analysis of your business (Figure 15.1). Recently, I delivered a few training programs for clients at two resorts. At both resorts, my clients were the only paying guests in the house. I am used to working in busy buildings with lots of activity in other rooms around me. Now, I had the undivided attention of the conference staff at both facilities. I learned that bookings for conferences and training sessions were way down from last year. At one resort, visits by tourists had almost completely stopped. I cannot pretend to be a marketing expert for the tourism industry, but, using the SCCIO analysis, here's how an example might play out, using the tourism industry.

Business Sector: Tourism and Hospitality

Situation: Low bookings for business conferences, group meetings, and training events, plus low bookings for individual guests. Hotel facilities are being underutilized; facilities look empty. Staff morale is low. Visitors ask, "Why is no one here?"

Causes: Reduced client budget for such events, fewer companies still in business, lack of time set aside for training or planning due to economic uncertainty, increased individual anxiety about discretionary spending, reduced willingness to travel. Not enough people are spending. Fear.

Consequences: Costs climbing faster than revenues, some facilities may not survive, staff may lose their jobs, needed renovations or other changes will be delayed, spin-off businesses may be hurt (suppliers of restaurants, cleaning services, uniform services, security, etc.), Fear will be contagious and self-fulfilling; business will continue to be bad.

Implications: Fewer facilities will be in business next year, more staff will be available, management must maintain a small cadre of good staff to offer quality services as the economic recovery builds, prospective clients will have fewer choices for destinations, facilities will have to cut costs, create more attractive pricing and innovative packaging, find new markets, offer new services, reinvent their businesses.

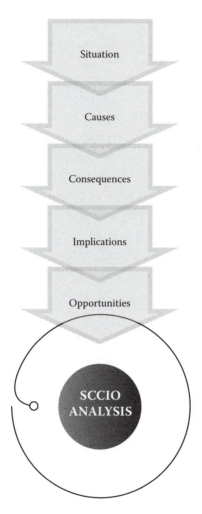

FIGURE 15.1
5 Components of a SCCIO Analysis.

Opportunities: Become a different, more creative business that attracts its own clients; create better, more flexible packages to appeal to a broader range of clients; strengthen your position as a key player in regional markets able to compete for new business as the economy rebounds; find new ways to control waste and cut costs; develop cross-functional staff to work in a wide range of departments; involve staff in finding better ways to design and run operations; research new themes and branding to become a niche destination for specific clients; research new technology as a means of driving more business; try Internet marketing in other countries; don't just be a *Survivor*, be a *Thriver*.

MANAGE YOUR PEOPLE WELL

- TELL THEM THE TRUTH. You may be surprised to learn how eager they are to find out more about the financial health of your business.
- GET CONTROL OF THE RUMOR MILL. Communicate often and maintain an open-door policy. Rumors flourish in a vacuum.
- CREATE THE PERCEPTION THAT YOU ARE ALL IN THIS TOGETHER and that now, more than ever, everyone needs to be a team player.
- SHOW YOUR APPRECIATION for their efforts. Say "thank you" and mean it. Acknowledge that times are tough and that you really appreciate their work.
- ACKNOWLEDGE THAT THEY ARE DEALING with more job insecurity, a higher cost of living, and stagnant salary levels. Show them that you care.
- IF LAYOFFS ARE NECESSARY, do them all at once. Do not cut a bit now, a bit more in a few months, and still more next summer. If the cuts are done all at once, then people will not be waiting for the next shoe to drop.
- BE PREPARED TO DEAL WITH MORE STRESSED-OUT REACTIONS at work. Encourage people not to take out their problems on their co-workers; let them know of community resources they can access for counseling or advice if required.

Of course the current downturn creates huge business problems. You need to find a way to see the opportunities that are also on offer. They are out there, but you have to look for them.

HOW TO MANAGE DURING TOUGH TIMES

- DO A SCCIO ANALYSIS of your business (Situation, Causes, Consequences, Implications, Opportunities).
- MANAGE YOUR PEOPLE WELL by telling the truth, dispelling rumors, emphasizing teamwork, showing appreciation, and acknowledging financial pain.
- IF LAYOFFS ARE NECESSARY, do them all at once.

16

Viral Leadership:
How to Create Positive Change

Ellen, a business owner in the service sector with a staff of 40 employees, was overwhelmed by my question: "What could you do to create momentum about the changes you want to see around here without spending money or taking people away from their jobs?"

The two of us were near the end of a long meeting; Ellen wanted to know what she could do to improve her bottom line. The issues of falling revenue, customer dissatisfaction, aggressive competition, and poor teamwork were welded together like sheet metal in a train wreck. She needed to turn her business results around in a hurry. The recession had taken a large bite out of her revenue, formerly loyal customers were defecting, and new competitors were slashing prices by offering a minimal service model. Falling morale, fear of job loss, and a lack of cooperation made coming to work each day an act of sheer willpower. Of course, Ellen had no budget for incentives, coaching, or training seminars for her staff.

She had reached the limit of what is possible through command-and-control leadership. Ellen had issued memos, talked to staff, disciplined poor performers, and tried to recognize her stars. These are some of the right actions, to be sure, but she did not get the right results. When she said in frustration, "I've done all I can. If we are going to get through this, people will have to do it for themselves." I almost burst into applause. I told her that this insight was the first stage of climbing up the mountain of trust. Leaders can only point the way. Employees have to climb that mountain themselves. But how?

For leaders like Ellen, Viral Leadership can be a game changer. Imagine an organization in which a culture of high trust, great teamwork, and excellence in execution spreads to every employee, seemingly by itself, just as a flu virus replicates itself and spreads through a community. Viruses can spread through host organisms with great speed. Unless a virus encounters antiviral factors, such as a healthy immune system or a vaccine, it can easily take advantage of those with weak defenses. Flip this scenario over on its positive side: A strong virus can quickly overcome many obstacles and be very successful.

The expression "going viral" was first used in the mid-1990s to describe the phenomenon of viral marketing via the Internet. Trends, opinions, expressions, and fashions seemed to spread spontaneously around the globe. Today, many large corporations go to great lengths to promote their products and services through viral marketing. In his 2008 U.S. presidential campaign, Barack Obama used viral marketing via social networking sites to raise more funds than John McCain, who stuck to traditional fundraising appearances and media campaigns. Obama's core message "Yes We Can" was everywhere, mostly as free messages in the news and on the Web via social media. Today "going viral" means "spreading spontaneously."

But, how exactly can a culture of high trust, great teamwork, and excellence in execution spread to every employee, especially without massive incentives, focus groups, training, and logistical support?

In his book, *Viral Change* (meetingminds, 2008) (http://www.viralchange.net/), Dr. Leandro Herrero tells the story of a new executive who made a series of relatively informal comments to staff members on where their business should be heading. These casual comments generated an avalanche of changes that ultimately improved the organization. When Herrero interviewed this executive a few months later and congratulated him for what seemed like a "good change management program," his reply was: "What change program?"

Of course, many leaders could make the same announcements and get nowhere. How can one leader make a few casual remarks that lead to a significant change, while other leaders saying the same things are ignored?

Viral leadership works best in an atmosphere of shared meaningful values, high trust, and deeply shared commitments. Everyone must be on the same page. How can you get them there?

Shared Meaningful Values: Leaders must identify values that are very important to their people. Then they must not only share those values, they must live by those values. What is a meaningful value? "Keeping my job," "getting a promotion," or "staying in business" simply does not cut it. These values, while good as far as they go, do not go deep enough into what matters most to us.

A few years ago a friend of mine was conducting a Vision–Mission–Values seminar for the sales department of a major cell phone provider. A cynical sales rep raised the following objection: "C'mon," said the cynic, "What's with all this values hokum? I don't sell values, I just sell phones." My friend, a New Yorker, paused for a moment, then said, "On the morning of 9/11, families with loved ones on the doomed planes and in the twin towers said good-bye to each other over the products you sell. ... Think about it. You sell more than phones."

A cell phone is not only a piece of communication hardware, it is also a means of bringing loved ones together over great distances. These phones can carry our hearts from just about anywhere on the globe.

The goods and services we provide hold our families and our society together. All of us, as a metaphor, "sell more than phones." Whatever the product or service, it can tap into a higher value than merely providing someone with a job. *Leaders can only discover these higher values by getting to know their employees.*

High Trust: Someone who truly practices viral leadership must go a lot farther than merely talking about trust. Leaders must *demonstrate trust* through their own behavior every day. It takes a significant level of personal courage to believe in the best side of people and to have faith in their better angels. To get the best out of people we must believe in the best sides of their characters. Leaders must encourage possibilities as well as set limits and enforce consequences.

Deeply Shared Commitment: Leaders need to demonstrate personal commitment to their employees' success. Again, *the messages that people will really hear will not be delivered by words, but by deeds.* Actions speak with much more authenticity than words. If, for example, a core belief of your organization is that "Our People Are Our Greatest Resource," then you must find a way to demonstrate that belief and not just talk about it. That might mean listening to employee concerns and acting on them instead of just gathering feedback for an elusive tomorrow that never comes.

Viral leadership depends on many additional factors beyond shared values, high trust, and deep commitment. Extraordinary communication skills and high levels of emotional intelligence also are essential, not only for designated leaders, but across the entire organization. When you can truly say that your people are helping each other become all they can be, then you are beginning to really foster Viral Leadership. "Of a truly great leader," Lao Tzu (ancient Chinese philosopher) reminds us, "the people will say: 'We *did this ourselves*.'"

**HOW TO CREATE POSITIVE CHANGE
THAT SPREADS BY ITSELF**

- FIND OUT THE SHARED MEANINGFUL VALUES that really matter to your people.
- DEMONSTRATE HIGH TRUST in your people; believe in their better angels.
- SHOW THAT YOU ARE COMMITTED to your peoples' collective and individual success.
- REMEMBER LAO TZU: "Of a truly great leader, the people will say, 'We did this ourselves'."

17

How Can You Learn to Make Better Business Decisions?

As a leader, you are paid to make business decisions every day. Or more accurately, you are paid to make the *right* decisions (Figure 17.1). What separates a great decision maker from the rest of us? Can you learn to make better business decisions?

Defend against Personal Biases: You need to challenge your own thinking. Invite dissent and disagreement. Encourage dialogue and debate. The investment billionaire Warren Buffet, the legendary "Oracle of Omaha," defends against his own biases in making large acquisitions by appointing an "advocate against the deal," who is well rewarded only if the deal does not go through. However, if you punish dissent and opposing opinions, you will immediately create a climate of false harmony. Your people will tell you only what you want to hear. Remember, you want honest feedback. That includes the feedback you may not like to hear, but that you really need to hear. So don't shoot the messenger.

Become Aware of Your Emotional Baggage: If you are considering buying an additional piece of business equipment, or enlarging your operation, you may find yourself being reluctant to take the plunge because of stored negative emotions from previous experiences. Are these past experiences relevant? You need to be able to learn from the past, of course, but you also need to let go of the past. A business investment failure of a few years ago may still color your judgment. Conduct a thorough postmortem, learn from your mistakes, and move on. Or seek objective advice from someone who knows your business, but does not share your emotional history with a previous failure.

Conduct a Premortem: When facing your business decision, ask yourself: "What could go wrong and what could I do to prevent this failure

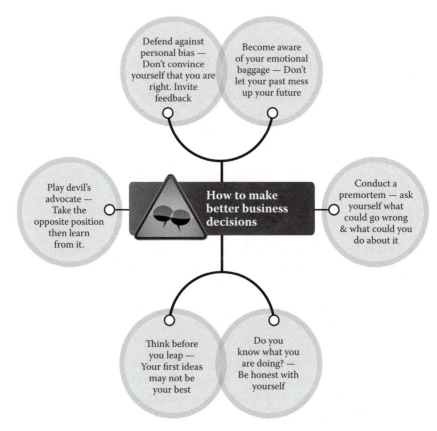

FIGURE 17.1
6 Ways to Make Better Business Decisions.

from happening?" Develop two lists: a list of solid facts and a list of uncertainties. Then ask yourself if you have the background to make good judgments about these uncertainties. A premortem is a technique used in scenario planning in the military and in large corporations. Decision makers need to ask themselves: "What could go wrong and what could we do to prevent it?" Other useful questions include: "What is the worst possible thing that could happen as a result of this decision and how could I respond to that outcome?" "What changes (in finances, technology, personnel, government regulations, customers, general business environment) could have a significant impact on the outcome of this decision?" And, "What could really go *spectacularly right* with this decision if I take the right actions now?" Are your dreams large enough?

Do You Know What You Are Doing? Have you made this sort of decision before? Do you have an appropriate bank of remembered experiences to draw upon for comparison or are you making it up as you go? Good computer analysts do not always make good marketing decisions and vice versa. Be fully honest with yourself about your business experience and abilities. It is far wiser to admit that you are in over your head in terms of risk management and expertise than to proudly defy the odds, then fall flat on your face.

Think before You Leap: Avoid "shoot from the hip" decisions when the consequences are significant. Do not become attached to the first idea that comes to mind. Don't go with that first idea. Give yourself the time to reflect. Ask for help.

Play Devil's Advocate: Once you land on a decision, make the opposite decision, then try to justify that decision. This oppositional thinking can trigger a much richer and better quality of decision.

By all means, listen to those gut feelings. Listen to your intuition. However, take the time to think through your own decision-making behavior.

LEARN TO MAKE BETTER DECISIONS

- DEFEND AGAINST PERSONAL BIASES: Don't convince yourself that you are right. Invite feedback.
- BECOME AWARE OF YOUR EMOTIONAL BAGGAGE: Don't let your past mess up your future.
- CONDUCT A PREMORTEM: Ask yourself what could go wrong and what you would do about it.
- DO YOU KNOW WHAT YOU ARE DOING? Be honest with yourself.
- THINK BEFORE YOU LEAP: Your first idea may not be your best.
- PLAY DEVIL'S ADVOCATE: Take the opposite position, then learn from it.

Section III

Communication

18

How Designed Conversations Get Winning Results

Any manager knows how difficult it can be to have an effective conversation with an employee about sensitive performance issues. Barriers to listening automatically spring up. Defensiveness can go through the roof. After such a conversation, employees can feel mistrusted and harassed while managers can feel that the message just didn't get through. When such conversations go off the rails, the damage can poison a workplace and hurt business results. Consequently, too many managers back off from these conversations. They may think, "Sure, I should deal with that issue, but it just isn't worth the risk." They learn to look the other way. Avoidance has never been a pathway to business or personal excellence, but we do it every day.

Performance issues aren't the only challenging conversation topic that can go wrong. Think of negotiations, budget discussions, or operational problem-solving sessions. The unfortunate reality is that many managers often are placed in situations where they need to have important conversations for which they lack the necessary communication skills. They need to have a designed conversation.

WHAT IS A DESIGNED CONVERSATION?

A designed conversation is a structured process for enabling all participants to fully engage and realize their goals. If a designed conversation is managed well, there will be clarity and agreement at the end.

A designed conversation involves engaging the services of a third person who can be viewed as being neutral by all parties. This third person, or coach, has the responsibility for designing a process for the conversation then staying in the room to ensure that the process is followed and the conversation gets to where it needs to go. This coach holds full responsibility for the process, but must remain neutral on the content once the conversation gets underway.

This is not to say that the coach must remain neutral about the goals of the conversation. Take the example of the performance review. Suppose a manager must address a difficult performance issue with an important employee. The coach for this designed conversation needs to support the goal of helping the manager and the employee understand the situation and agree on a solution. Part of the preparation for this conversation would be to ensure that both parties come to the table supporting the goals for the conversation.

A designed conversation is a small group process. While it would be pointless to set an arbitrary maximum number of participants, the fewer the participants, the greater the likelihood of a successful outcome. Suppose you are a sales manager and two of your reps are not sharing leads between their territories. A skilled coach could design a process for a designed conversation between the three of you. Now suppose you, as the manager, wanted to have a similar conversation with your entire staff of 30. There are many more variables at play. A coach might find it more difficult to design a process that would address all of these variables, and the chances of arriving at a desired outcome might be less than if the conversation only involved three people.

EIGHT STEPS TO RUNNING A DESIGNED CONVERSATION

The coach

- **Meets with the Sponsor** to develop goals for the conversation;
- **Clarifies the Facts** and the perceptions of the situation with the sponsor;
- **Identifies** any areas of potential conflict;
- **Meets with All Parties** individually prior to the conversation to establish goals, agree upon facts, and to share a process for reaching agreement;

- **Aligns Individual Goals** and perception of facts with overall goals;
- **Designs a Process** for the conversation, including problem-solving strategies;
- **Shares Goals, Facts, Perceptions,** and the conversation process with all participants; and
- **Establishes Ground Rules**, facilitates, then evaluates the designed conversation.

A designed conversation is not a magic cure for all communication problems. However, it can be a powerful aid to helping people understand each other.

TO RUN A DESIGNED CONVERSATION

- FIND a neutral coach
- AGREE ON facts, perceptions, and goals
- ALIGN individual perceptions and goals
- AGREE ON PROCESS for handling disputes
- STAY FOCUSED on outcomes
- RESPECT the process
- TRUST each other

19

How to Get More Great Ideas at Work

Business owners, managers, and entrepreneurs wrestle with seemingly intractable problems all the time. The best of them realize that the solutions to their problems often can be found among those who know their businesses best: their employees. Their challenge is: "How can I get my people to help solve these problems that are holding us back?"

Developing innovative ideas that can overcome huge business challenges is not like implementing a straightforward action plan. You can't just flip a switch and have a deluge of great ideas land in your inbox. As Ivy Ross, executive vice president, product design and development, at Old Navy Clothing says:

> Innovation is not a process. It's creating an environment that helps teams of people quickly build trust and relationships. Then, people have the right framework to create.

Ross is only half right. It's true that you have to create an atmosphere that helps people become willing to trust each other with new ideas. People need to feel excited by the possibility of coming up with a fresh way to overcome tough challenges. The old "suggestion box" with a free lunch for the winning idea each month won't cut it. Why? Because it focuses only on individual ideas, not the synergy of many minds. And, it is not exactly exciting.

Of course, teams need to have a supportive environment to come up with innovative ideas. Innovation is not just a process. Innovation must *include* an effective process as well for converting random ideas into effective actions. Most people don't know how to generate innovative ideas. Shutting a group in a room with a good supply of coffee, muffins, and flip charts may not result in any profitable ideas. Everyone thinks innovation is just a matter of brainstorming. They think they know how to brainstorm. Often they don't.

Last summer, I was asked to help a small manufacturer come up with ideas for improving the company's financial situation. Sales were down. They had already squeezed all the cost savings they could out of their operation. The wolves were at the door.

First of all we precisely identified the challenge. Next, we assessed the current situation. Then, we developed a challenge statement. This challenge statement must be framed positively. The challenge statement this client came up with was: "We have found new markets, domestic and offshore, that have become highly profitable."

Next, we created an environment for a brainstorming session using 10 guidelines:

1. Everyone has to participate (no passive spectators).
2. Everyone must respect each other's intentions.
3. No sarcasm or put-downs.
4. Say what you need to say, but say it as concisely as you can.
5. Don't fly solo, stay with the team.
6. Be an active listener and stay positive.
7. Each person must take responsibility to be heard.
8. Don't censor yourself and don't criticize others.
9. No judgments; have fun.
10. Stay positive no matter what.

Then, we created descriptions of what the ideal future might look like. I posted this question at the front of the room: "Wouldn't it be great if … ?" The ideal future question we came up with was: "Wouldn't it be great if we repositioned our products so we could market them to new customers, no matter where they are?" We came up with a number of ideas about what we could do to make this happen, then we grouped them under a few key themes. Next, we asked the question—"How could we do this?"—of each of these key themes. These statements became our Opportunity Ideas. We posted all these Opportunity Ideas, discussed their pros and cons, then select the Best Opportunity.

Now, we were ready to brainstorm. How could we convert our Best Opportunity into a reality? We generated flip charts full of ideas on Post-it° notes. People started to complain that they had run out of steam. I pressed them to keep going, no matter how far-fetched their ideas became. When they finally refused to go any farther, we returned to our Best Opportunity and applied three Innovation Tools:

1. **Making Comparison:** Compares the Best Opportunity to another unrelated system that appears to have a few parallel qualities.
2. **Forcing Associations:** Use a series of provocative words or pictures as triggers for new ideas.
3. **Switching Assumptions:** Reverse the usual assumptions behind a Best Opportunity.

It turned out that switching assumptions led to an idea that eventually proved to be our breakthrough. Most manufacturers assume that their clients want to buy products that will last forever. Our new assumption was: "Suppose clients are looking for a product that they can break down and recycle after they are finished with it, as opposed to storing for future use?"

At last, we were ready to test our ideas against two criteria: Profit Potential and Capability (are we able/capable of doing this?). We took all the ideas we generated in Brainstorming and the ones that came out of using the Innovation Tools and tested them against Profit and Capability. We used a simple quadrant grid set out on the floor of a large, empty room with masking tape. Everyone joined in the chaotic shuffling of Post-it notes from quadrant to quadrant until we reached consensus.

Next, we focused on the ideas that had been placed in the upper-right quadrant (High Profit Potential and High Capability). I gave everyone three votes in the form of three colored stickers. The idea with the most stickers was our winner.

This manufacturer arrived at a solution to their problem that was far beyond their expectations. They got there because they were willing to learn how to develop innovative ideas. Innovation cannot be taught. However, with the right attitudes and the right tools, leaders can help make it happen.

HOW TO GENERATE INNOVATIVE IDEAS AT WORK

- DEVELOP A CHALLENGE Statement
- FOLLOW THE 10 brainstorming rules
- SELECT AN IDEAL FUTURE or Best Opportunity Idea
- BRAINSTORM TO COMPLETE FATIGUE, then review Best Opportunity Idea
- MAKE COMPARISONS, force associations, and switch assumptions using Best Opportunity Idea
- USE A PROFIT POTENTIAL AND CAPABILITY MATRIX to select winning ideas

20

Are Intergenerational Conflicts Hurting Your Business?

Recently, I was asked to go into a midsized technology firm and deliver a conference presentation on intergenerational communication. When I started asking questions to find out exactly what they meant and what they wanted from me, a familiar pattern emerged: Employees of different generations, working together, but not understanding each other and assuming the worst about each other's intentions.

The business impact of this lack of trust and poor communication was that the firm's clients were experiencing delays in product delivery and poor customer service. Some senior managers were defaulting to very authoritarian leadership styles with a predictable outcome: A steady trickle of young, promising employees were quitting, taking their knowledge and skill sets elsewhere.

Suspicion and mistrust between the generations have been facts of life at work since ancient times. Why are these issues such a big deal today? What makes the current situation absolutely unprecedented in history is the explosive development of new technology, the rapid expansion of the globalized economy, both combined with a highly unusual demographic pattern in the workforce.

Simply, the presence of four generations in the workplace at the same time creates tension. The veterans, born before 1946, are still around, although in diminishing numbers as they retire; many of the "boomers," born between 1946 and 1964, are now in senior leadership roles. Generation X, born between 1965 and 1979 are taking over from the boomers and veterans, while generation Y, born since 1980, is beginning to move into the workforce in large numbers. Demographers disagree about exactly which years represent the specific generations and their labels, but the consensus seems to be that prevailing generational characteristics shift roughly every 20 years.

Interesting stuff, but I knew my audience would not respond well to a lecture on demographics and economics. They had a real-time business problem that would not be improved by a dry, academic presentation on theories and trends.

My client's employees needed to understand the basic differences in how the generations see the world, so I put together a slide presentation that covered the origins of the differences between each generation (different historical contexts and extremely different child-rearing norms), the impact of these differences on lifestyle and, most importantly, on work style. This presentation concluded with a module on how to communicate with each generation and how to manage Gen Y.

However, I knew I couldn't get away with just subjecting my audience to two hours of PowerPoint˚. They needed much more than my research and my imagined—and certainly limited—expertise. So, I asked for two volunteers from each of the generations to join a panel at the front of the room for a discussion. I worked this panel through a number of discussion questions, such as:

- "What don't you understand about other generations who work with you and how they communicate?"
- "What do other generations need to understand about your generation?"
- "If you could fix only two things about other generations and how they communicate, what would those two things be?"

Out of this discussion, the group came up with the following points to remember when dealing with each of the generations:

When managing boomers:
- **Give** them the big picture.
- **Tell** them they can make an important contribution to the success of the team.
- **Recognize** them by providing recognition and increased profile.
- **Individual** boomers may prefer phone or face-to-face over email.
- **Acknowledge** their experience and recognize that they have "paid their dues."

When managing Gen Xs:

- **Appeal** to their sense of personal loyalty, not loyalty to the organization.
- **Give** them a way to "buy into" a project rather than bark orders at them.
- **They** respond better to short-term objectives rather than long-term goals.
- **Tell** them what needs to be done, but not how.
- **Give** them multiple tasks, but allow them to set priorities.
- **Remember** that they respond best to informal recognition rather than formal acknowledgment, such as a plaque on the wall.

When managing Gen Ys:

- **Give** them opportunities for continuous learning and building skills.
- **Find Out** their goals, then explain how these goals fit into your organization's big picture.
- **Be** more of a coach, less of a boss.
- **Communicate** with them via informal hallway conversations and email.
- **Give** lots of feedback and recognition.
- **Do not** rant or humiliate when giving constructive feedback.
- **Remember** that they will not tolerate inauthentic leadership. Instead they will leave.
- **Ask** rather than tell.
- **Remember** that for them, the teambuilding rituals so important to boomers are simply a waste of time.
- **Keep in Mind** that they work to support their lifestyles outside of work, not out of any sense of loyalty to a larger entity, such as their employer.
- **Anticipate** their needs for work/life balance, authentic leadership, and continuous learning.
- **Give** them a clear picture of their career future and how they can advance.
- **Ask** them for input on decisions that will affect them.
- **Avoid** criticisms and reprimands; instead point out errors, remain emotionally neutral, offer positive alternative approaches, and make plans for improvement immediately.

At the end of the discussion, I invited the audience to respond to a few of the questions put to the panel by writing their answers on Post-it® notes without their names then sticking these notes up on the wall for all to read. Apart from one snarling remark ("Send Gen Y off to Boot Camp"), the overwhelming tone of the answers was of tolerance and a need to accept people of all ages for who they are and appreciate what they can bring to their work.

The job of today's managers, business owners, and entrepreneurs is to find ways to leverage each generation's distinctive talents.

21

How to Run Effective Meetings

Once I had to watch the executive team of a large pharmaceutical firm slog through a two-hour business update meeting. Sounds dull? If you weren't a shareholder in this company, then, yes, that meeting would have been painfully dull. However, if you were a shareholder, then you would have been totally infuriated. These executives bickered, didn't listen to each other, failed to make any decisions, avoided key issues, got lost on tangents, and wasted a lot of time. As I was sitting next to their Human Resources director, I asked her if she could give me a rough idea of the annual salary levels represented around that executive table. I did the mental math and came up with a per-hour salary cost for that team of close to $1,500. Did the company's shareholders get $3,000 of value for that meeting? I'm sure that they would have unanimously voted "no!"

Over the years I have noticed that when it comes to running effective meetings, many otherwise extremely talented professionals can be completely inept. I have watched teams of lawyers, investment bankers, accountants, project teams of software engineers, corporate executives, sales and marketing MBAs, doctors, and academics flounder through meetings, wasting a lot of time, to say nothing of their considerable salary costs.

The basic rules for conducting effective meetings are taught in high school. Then, why is it that very competent adults at the top of their careers cannot do the same? Frustrated by this realization, I developed a cynical list of negative meeting rules for a management teambuilding program. My hope was that in the midst of their laughter, my participants would recognize some of their own dysfunctional meeting behaviors. I asked everyone to pick out their favorite bad rule and give an example. The ensuing discussion helped them realize how they could improve their game.

Herewith are my **Top Ten Rules for Running Really Bad Meetings**. If you find yourself cringing in self-recognition, then read the refresher that follows:

Rule #1: Don't do any planning beforehand. Don't think about what you want to get out of the meeting as a result. Just get everyone together and let it all hang out.

Rule #2: Waste peoples' time. Use a lot of the meeting for transmitting information using one-way communication. Discourage dialogue. People might learn something if they discuss an issue.

Rule #3: When discussions start, let them run in whatever direction they choose. Encourage free speech and avoid all structure.

Rule #4: Encourage people to come unprepared to make decisions, then avoid using an agenda or any time limits so that you run out of time to make decisions. That way you can call another meeting next week.

Rule #5: Don't pay attention to the clock and let the talkative people do all the talking. The quiet people probably have nothing to say, so let them daydream or doodle away their time.

Rule #6: Let people repeat themselves and get off track. Don't press anyone to make a decision.

Rule #7: Don't explain in advance or solicit buy-in for any decision-making or conflict-resolution processes that you would like to use. Let everything hit the fan. Then when people start screaming, ask them to quiet down and present your brilliant ideas.

Rule #8: Don't assign any roles in advance that might help get something done. Roles to avoid assigning include Facilitator, Chairperson, Timer, Minute Taker.

Rule #9: Let the ranters rant. Discourage listening. Encourage several people to speak at the same time. Never ask anyone to summarize what they just heard or paraphrase to confirm understanding.

Rule #10: Above all, never, never do any process checks, wherein you act as a facilitator and summarize what has happened in the meeting so far, give feedback, then invite suggestions as to how to proceed next so that decisions get made and work gets done.

A QUICK REFRESHER:
HOW TO RUN EFFECTIVE MEETINGS

Stephen Covey, acclaimed author of the perennially popular bestselling book, *The 7 Habits of Highly Effective People* (Rosetta Books, 2009) (online at: https://www.stephencovey.com/7habits/7habits.php/), advised to always "begin with the end in mind." This is particularly true of team meetings. When a team leader decides that a meeting is necessary, this person should always reflect on a question, such as: "What do I want to achieve at the end of this meeting? What outcomes do I want?"

Running effective meetings should be a widely practiced competency in all organizations. Unfortunately, this is not always the case. Any team, with a bit of foresight, can set up and run highly effective meetings. Follow these steps to ensure that your team gets the results it wants out of a meeting.

Before the Meeting

Create and Circulate an Agenda to all attendees beforehand. A detailed agenda includes a list of topics, what needs to be achieved per topic, a time limit for each topic, name of the person presenting or facilitating the topic, details of the process to be used for each discussion.

Design Agenda Items so that they result in an outcome.

Plan Your Facilitation Process for the meeting. If an agenda item will require some discussion, then develop a process for managing that discussion and making a decision. Estimate the amount of time this discussion will take and build that time slot into your agenda.

Do Not Design Items That Are Merely for Sharing Information. Sharing information can be done offline.

Clearly Specify What Preparation You Expect of your attendees (pre-reading or other research).

Circulate a List of who will be at your meeting.

Set Your Logistics up So That Everyone Can Fully Participate (time of day, location, parking, refreshments, etc.).

Distribute Presentation and Reporting Duties along with note-taking beforehand.

During the Meeting

Be Sure That Someone Will Play the Role of Facilitator. The Facilitator keeps control of the process (how people are working together as a team during the meeting), not the content or agenda items. The Chairperson is not necessarily the best choice for Facilitator. The Chairperson should be in charge of the content of the meeting. The Facilitator should stick to managing the process of how people work together during the meeting. Frequently, an outsider, even someone from outside the organization, can be the best choice for Facilitator.

Stay on Track. Allot a set number of minutes per agenda item beforehand then stick to your agenda.

Establish and Get Buy-In on norms for behavior.

Manage Participation so that people use appropriate discussion skills.

Be Explicit about the Process to be followed for making decisions and resolving conflicts.

Take Notes or Minutes of the meeting, particularly any decisions. Be sure to capture who will follow up, what actions are to be taken, and by when.

Create Buy-In to these processes by asking people to contribute their own process ideas.

Reach Closure on each agenda item or at least identify next steps that could include appointing a working group, putting the item on the agenda of the next meeting, or postponing the item until necessary research and preparations have been completed.

Solicit Opinion from everyone present, not only the most vocal.

Do Not Allow Individuals or cliques to dominate.

Be Explicit about Next Steps, including details and agenda for the next meeting.

Create Ownership wherever possible by inviting participation in decisions for actions and next steps.

At the end of the meeting conduct an evaluation of what people liked about the meeting, what they didn't like, and what should be done to correct any problems. Or use Keep–Stop–Start headings.

Afterward

Circulate Minutes along with preparation materials for your next meeting.

Encourage People to Read the Minutes and make corrections next time.

Developing a business culture of effective meetings takes time and discipline, particularly if your organization has a history of bad meetings. Be patient as you implement these changes in meeting behaviors. Bad habits take time to change. However, the eventual outcome—effective meetings that get results—will repay your initial investment of time and energy many times over.

22

The Basics of Giving and Receiving Feedback

Everyone needs to be able to give and receive constructive feedback in a clear, positive way. In an ideal organization, feedback should be seen as a gift, not an attack. Why a gift?

Feedback tells us if our actions are effective. Feedback shows us how our actions are helping us get the results we want. Feedback also tells us how our actions influence others.

Suppose your actions are getting in the way of your ability to get the results you want. Suppose these same actions also are causing unnecessary hardship and feelings of mistrust for others.

Without receiving any feedback, you would never know, until perhaps it was too late, that your actions were not getting the results you wanted. You would never know that your actions were creating difficulties for others or hurting your own career.

Without receiving feedback, you are in danger of creating so much damage to your career and your work relationships that you could risk losing your job. You also risk being the cause of a low productivity for others.

Years ago, I worked for a wilderness program in Northern Ontario as a canoe instructor. We took great pride in our knowledge of the bush and our abilities to safely guide our students down whitewater rapids, over rugged portages, and across huge open lakes. We especially took pride in our paddling skills.

One spring during staff training, a visitor from another program suggested that we bring in a professional coach to improve our paddling techniques. The response from my peers was instant and defiant. "Bring in a paddling coach? We've grown up in canoes! Paddling is natural to us, like learning to walk. Who could possibly teach us anything about paddling a canoe?"

So, it is with coaching managers and their staff on feedback. Everyone thinks that giving feedback is as natural as breathing. "You just bring 'em into your office, shut the door, look 'em in the eye and put all your cards on the table. What else could there possibly be to learn about something so straightforward?"

Like paddling a canoe, giving and receiving feedback is a skill set that can be vastly improved with training and practice. Here are the basics. Unfortunately, when people hear the word *basics*, they assume that the topic to be addressed must be simple to understand and that grasping the following points will indeed be easy. The challenge comes in putting these guidelines into practice.

GIVING FEEDBACK

Feedback should be given frequently enough so that it is seen as *a normal part of individual or team working experience.*

Regular feedback between individuals and teams can focus on preventative maintenance, acknowledging positive performance, and addressing minor irritations before they become major problems.

If feedback is shared between individuals or teams on a regular basis, then everyone knows where they stand. They *build up enough trust* in each other to be able to accept and value what they are told.

Feedback also should be positive as well as constructive.

While the technique of always starting with positive feedback before giving constructive feedback can seem manipulative to many people, try to strike a balance between positive and constructive feedback.

Try to catch people doing something right, as well as keeping track of when they do something that you want them to stop or change.

How to Give Constructive Feedback

- Take the time to **Find Some Privacy**. Only the people who need to hear the feedback should be present. If the feedback involves one team speaking to another team, then the two teams need to get together in private.

- **Be Specific and Use Examples**. Instead of making blanket judgments, as in: "Whenever you work with our team, our people get frustrated and angry because of your demands," say: *"When you ask us to change our procedures for clients without offering an explanation or without asking for our ideas, people sometimes ignore or resist your requests. Did you know that? Is that something you can change?"*
- **Be Specific about the Changes in Performance or Behavior That You Would Like to See.** To continue with the example in the above bullet point: *"When you ask us to change our procedures for clients without explaining why or asking for our ideas, people sometimes ignore you or resist your requests. Did you know that? Is that something you can change? For example, why not ask Jack about technical issues and how he thinks they could be improved before telling us that you want us to do it your way?"*
- **Do Not Rant or Get Off Topic.** Plan the feedback you want to give in advance then stick to your own agenda. Do not bring up additional issues unless they are related. Do not bundle issues together as in: *"And while we are on the subject of your punctuality, I have to remind you to please not park in the Visitor Parking Area when you arrive late and dash into the plant."* The fact that someone is not parking in the designated place and inconveniencing visitors is a separate issue. Don't let the issues pile up so that they have to be addressed all at once. No one wants to listen to an avalanche of constructive feedback on a wide range of topics.
- **Stay Constructive. Talk about Actions, Attitudes, and Behaviors**, not personality issues. People can change what they do, what they think, and how they behave. They cannot change your perception of their personalities.
- **Encourage Dialogue.** Ask: "Does this make sense to you?" or "How do you see this situation? Am I being fair here?"
- **Don't Try to Establish Who Is Right or Wrong in This Dialogue**. Do not let dialogue slip into debate. Both parties need to accept the fact that feedback is based on perceptions as well as facts, and that each person's view must be accepted as being valid.

Practice Compassion. Remember that you are not trying to get revenge or payback for something that you perceive as being wrong. Your goal in giving feedback should be to help that person or team improve for his or her own good and for the benefit of the entire organization.

How to Receive Constructive Feedback

Remember, You Are Receiving a Gift. Say: *"thank you."* Far better that you hear constructive feedback directly from the people involved rather than through the rumor mill or, worse yet, not at all. People will give you feedback because they have faith in you, your desire to listen, and your willingness to change.

Do Not Fight against, Debate, or Argue with the Feedback. Remember Stephen Covey's words: *"Seek first to understand, not to be understood."* In other words, focus on understanding what you are being told rather than trying to make the other person understand what you mean.

Ask Clarifying Questions, such as: *"Can you give me an example?"* or *"What do you mean? Can you tell me what you want using different words?"* Show the person that you really want to understand what she or he is trying to tell you.

Practice Active Listening. Focus all your attention on the person who is talking to you.

Do Not Reject the Feedback simply because you don't want to listen to it. Few of us naturally want to find out about our weaknesses or shortcomings. Even if the person is wrong, it is her or his perception that has led to this feedback. You must honor that perception.

Acknowledge That Giving Constructive Feedback Calls for Courage. Do not underestimate the stress that the person giving you feedback may be experiencing during the conversation. Thank them for taking the risk to talk to you.

Receiving Feedback Can Be Painful. To feel discomfort or sadness during feedback is normal. Avoid blaming other people for your own feelings. After you have thanked the person or team for their feedback and the conversation ends, spend a few minutes reflecting on what you have heard. Decide how much of this feedback you need to accept as valid, then decide what you are going to do about it. If appropriate, get back to the person who gave you the feedback with your plans for change.

Giving feedback effectively will make you a better leader. Receiving feedback graciously will strengthen your relationships.

Section IV

Your Career

23

How to Manage Your Career

As you become immersed in the demands and challenges of being a frontline manager, it is all too easy to forget that you are the only person in your organization who can effectively manage your career. This may sound overly simplistic. Who else could possibly look out for your best interests? However, experience teaches us that many otherwise proactive, capable, and mature employees are held back by an assumption that someone else, someone on high in their organization, someone with a special concern for each employee's well-being is looking out for them. Perhaps this is a shadow of an early childhood belief that no matter where we were, a parent or other caring adult was there to look after us. This is a wonderful assumption for small children; this assumption gives kids the courage to take risks and to expand their horizons as they learn and grow.

But, as adults at work, this assumption can make us dangerously passive when it comes to taking initiative and advocating for ourselves. The reality is that there is no older, parent-like figure at work looking after your career. In most contemporary organizations, there is relatively little direct performance management. Chances are there is no one to tell you, day in and day out, how you are doing. Unless you actively seek out a career mentor, there may be no one who will give you feedback on your progress or encourage you to take on a new challenge. There is no one to tell you if a given position will enhance your future employability or lock you into a dead-end job with slim chances of promotion. There is no one to encourage you to take on new projects to expand your knowledge base or grow your skill sets. All these career management vacancies can only be filled by you.

By all means, seek advice from people you respect. Find a career mentor if you can. Talk with others facing the same challenges. Don't try to manage your career on your own, but never assume that someone else will do it for you.

Looking after your own career can mean different things to different people. Above all else, you must keep in touch with your own reactions to your work. Periodically ask yourself these questions:

- How well am I doing? Am I doing my best work? If not, why not?
- Does this work matter to me? Is this work important in any way to me, beyond getting a regular paycheck?
- Am I learning more, am I making progress, or am I just putting in time?
- Do I see a path forward to more fulfillment, more learning, and greater responsibility? Or am I in a dead-end job with no prospect of a change for the better?
- Is the pay adequate? Do I see any opportunity for improvement?
- Is the number of hours or shifts likely to remain the same or increase? Or are layoffs a possibility?
- Can I see myself doing the same job five years from now? If not, why not?
- Am I actively researching other options?

If the answers to five or more of these questions are negative, it might be time to start thinking about making a change—a new assignment, a lateral move, a promotion, another job doing similar work with a different organization, or a complete career shift. Or you could decide not to decide.

If you don't make a decision to change, thereby coasting in the hopes that your career will continue as it has so far, you might be in for an unpleasant surprise. When you let yourself drift, other people notice, including your manager. If you don't take charge of making the changes you want to see in your career, the changes might be made for you, and not always in your favor.

None of the foregoing is meant to suggest that there should not be times when you must take a job you don't like in order to earn a living. These compromises are occasionally an unavoidable part of the contemporary job market; you should not think badly of yourself for having to sometimes take such jobs for the money. So much depends upon your own level of awareness. If you can acknowledge that you are making a career compromise and can accept the conditions of that compromise and make your peace with these less-than-ideal conditions, then at least you are not deluding yourself. Having this level of awareness will permit you to keep searching for alternatives and not lose touch with your dreams. The degree to which you will be able to eventually get out of these uncomfortable compromises and change your career is dependent upon how proactive you allow yourself to be in creating the changes you want.

Many of us do not accept our full degree of agency over our careers. We think that our careers are predetermined by our background, our age, or our level of education. To some extent this is true. However, we don't always accept sufficient accountability for creating the results we want. Therefore, we drift. Sometimes we even sink to blaming others for our career malaise.

Much is made these days of pursing our passion at work and striving for complete career fulfillment. Yet, if we look about our communities and our social circles, we will notice that very few people wind up doing work that is their life passion. Therefore, are the rest of the populace total failures? Not at all. So, the art of making smart compromises becomes an essential career management skill.

Consider this example: Harry is a top-notch pianist. From an early age, he studied with master teachers and has an impressive repertoire. Great music is his passion; nothing gives him a more profound sense of being alive than performing his favorite Bach composition. However, by the age of 25, Harry had to accept the fact that, while he was gifted, he was not great, and neither he nor his family could afford the expensive training required to become a professional concert pianist. He had built up a significant student debt and he knew he had to start earning a living. A summer internship with an established publisher led to a job offer as a contract researcher on an education book series on the arts. Harry applied himself and was rewarded with promotions. Today, he is an assistant editor and works on projects that result in creative learning about the arts that appear in print and online for schools across the country.

Harry knew he was a good pianist, but he also knew that he was never going to be the next Van Cliburn. He was very, very good, but he was not truly great. He wanted a career as a concert pianist, but he knew he could not consistently perform at that level. So, he applied his passion for music to a rewarding publishing career. Did Harry compromise his passion for the sake of money, or did he make a smart decision that allowed him to apply his skills and talents and earn a decent living at the same time?

"Living your life's passion" can be an immensely alluring objective early in our careers. When we are young, we might harbor visions of ourselves winning Olympic medals, becoming brilliant inventors or entrepreneurs, creating a great film, winning a Nobel Prize for our scientific research, winning an Academy Award, becoming wealthy celebrities, or some other outstanding achievement. Author Malcolm Gladwell, in his best seller, *Outliers: The Story of Success* (Back Bay Books, 2008), contends that one of the factors, albeit not the only factor, in creating a brilliant career, is to put in 10,000 hours of practice.

While Gladwell's "10,000-hour rule" is a dangerous oversimplification of what it takes to be the best in any field, we need to question the assumption that being successful equals being the best. In the early 1960s, many young bands learned to play rock 'n roll. Only one of them became the Beatles. As Paul McCartney has observed, there were many, many bands in Liverpool that put in 10,000 hours of practice. Were all the others, therefore, abject failures?

The problem with Gladwell's book is that he links following a career passion with lots of practice as a means of attaining excellence. This emphasis on passion, effort, and excellence may be inspiring, but it is useful? Is it possible to have a fulfilling career without being at the very top of your field? Is it possible to be a successful amateur? If you conclude that the answer to this question is "yes," then the journey to becoming a successful amateur lies in cultivating the ability to make smart career compromises while at the same time striving to find what is most important to you in your career. Such a stance cannot be decried as "selling out," rather, it is a commitment to becoming all that you can be while at the same time adapting successfully to the world around you. Far from abandoning your own dreams, a smart compromise can be a way to make those dreams become your reality.

HOW TO MANAGE YOUR OWN CAREER

- TAKE FULL RESPONSIBILITY FOR YOUR OWN CAREER PROGRESS. Don't assume that good luck will find you. Create your own breaks.
- SEEK ADVICE AND MENTORS, BUT DO NOT WAIT FOR THESE SUPPORTERS TO FIND YOU. Be proactive. Don't play the "wait and see" game. Take charge of your own working life.
- HAVE REGULAR MEETINGS WITH YOURSELF TO ASK THE TOUGH CAREER QUESTIONS. Keep a journal, reflect on your work, and cultivate your own intuition.
- EXPLORE THE ART OF MAKING SMART COMPROMISES. If you cannot get 100 percent of what you want, how could you get 70 percent or 80 percent, and what would it take to be able to live with that outcome?
- DON'T ABANDON YOUR DREAMS. See how much you can get of what you really want in your career by making smart compromises.
- NEVER GIVE UP YOUR AMATEUR STATUS. Focus on doing what you really want to do rather than becoming the best in terms of status or position.

24

The Rise and Fall of a Dictator: A Leadership Case History

Roger ruled his school board with fear. As an ambitious teacher, then principal, then superintendent, he quickly rose through the ranks to become one of the youngest directors of education running a large public school district in the country. He was a strict disciplinarian who valued control above all else. Never afraid of a tough decision, he was ruthless in pursuit of his goals. His allies were rewarded with plum assignments and promotions, while opponents were crushed and driven off the board.

As a master strategist, Roger used his influence to create an atmosphere of subtle intimidation. He would meet privately with senior administrators before board meetings to establish priorities and work through tactics for managing the board's elected trustees, most of whom he dismissed as simpletons. His cheerful social manner and eager smile created a false impression of mutual trust and respect. Roger would ensure that his trustees only received information that supported his strategies; he applied relentless pressure on his superintendents to carefully filter all information released in these public meetings. Roger was in charge, and those who resisted his leadership were isolated and ultimately rendered ineffective.

Like most school boards outside of major urban areas, Roger's school district was experiencing a steady decrease in enrollment. The declining birthrate of the past few decades meant that fewer and fewer children were registering for school each fall. Only schools in areas of significant immigration were growing. Roger was faced with what most directors seek to avoid; namely, the unenviable task of closing schools.

School closures are never popular with the communities losing a facility. Students demonstrate and walk out of classes. Aggrieved parents flock to board meetings, clog meeting agendas with their impassioned presentations, and occasionally take a board to court in order to try to prevent a school closure. When a school board wants to close a school, the government dictates that a standard public relations process is used to ensure that the school community—parents, families, and students— accepts the consequences of the closure. School communities must be provided with a process that gives the illusion of participation in the closure decision, even though the conclusion is predetermined by the board and supported by the state government's Department of Education. Everyone must feel that their opinions have been considered, despite the fact that the closure decision in question has already been made behind closed doors.

For Roger, these set pieces of political theater were his element. He enjoyed structuring the process to ensure that the desired school was closed. He was not above altering meeting minutes and editing reports so that he always had supportive documentation available for the public. One consequence of Roger's management style was that board employees were often forced to alter facts and, in some cases, lie to the public. Because these administrative employees feared losing their jobs, they never dared to defy his orders.

Roger had decided that it was in the best interests of all the students in his district to keep Mapleville Regional Secondary School open; this school was a sprawling suburban high school with plunging enrollment. The board had sunk millions into this school; it had the most student spaces (1,600) and the best facilities for athletics and the sciences of all their high schools. Unfortunately, it was built during the 1960s, at a time when demographers were predicting a large population increase in the suburb where it was located. Over time, the suburb had transitioned from families with children to a growing retirement community. The only way the board could keep the school open was by closing smaller schools elsewhere and busing students. The Mapleville school had a student capacity of 1,600; its student population had dipped below 600. Government funding could be withdrawn for any school with enrollment below 75 percent of capacity. Mapleville Secondary was below 40 percent.

Roger's solution was to close a smaller, fully enrolled, historic downtown high school that hosted a very popular arts program, and bus the 750 downtown students out to Mapleville Secondary. Roger's stated rationale for this decision was that a large school with over 1,300 students could provide a wider selection of courses than the downtown school. He chose to ignore research that found the benefits of a wider selection of courses in large schools to be illusory. He also ignored the research that found that smaller schools get better learning outcomes than large schools, such as Mapleville Secondary. Moreover, he knew that he could not ask the Department of Education to fund the construction of a new high school on the other side of town if the old downtown school was permitted to remain open.

Roger cleverly manipulated the board's decision-making process during the meeting at which the trustees voted to close the downtown school. Despite a huge outcry from students and parents, the decision to close the downtown school easily passed. Roger smirked contently as weeping students and their angry parents filed out of the meeting. The Boardroom Wizard had won again.

Community resentment of the board's decision exploded. Students demonstrated, and parents petitioned the government to intervene. Stories opposing the board's decision appeared on television, in print, and on social media. Parents researched the feasibility of keeping both schools open by sharing the use of school facilities with appropriate community agencies, such as day care centers and libraries. Critics demanded to know why Roger had not approached the local community college or the university to share space at Mapleville Secondary. The implacable Department of Education maintained that the decision to close a given school was the sole responsibility of the local school board and, therefore, they could not intervene. Eventually the parents began raising funds to take the school board to court in order to seek an injunction. The late June court date came and again Roger performed brilliantly as he demonstrated that his board followed standard government procedure in closing the downtown school. The judge ruled in favor of the board and the case was dismissed.

The board's staff and trustees were greatly relieved, as were local government officials. The avalanche of angry letters and earnest delegations at board meetings with their suggestions for dealing with the low enrollment problem finally stopped. Business could return to normal. Roger was praised by supporters of Mapleville Secondary as being their savior; with the influx of the 750 downtown students in the coming fall, their school's future was assured. The board could get on with planning for the transition at Mapleville Secondary, as the staff prepared the building for the incoming downtown students in September. The downtown school was closed and prepared for sale to a condo developer in the fall.

Roger should have had a relaxing vacation that summer. Instead, he was bothered by rumors concerning the number of downtown students who would actually register for school at Mapleville Secondary in September. In order to keep government funding for the school, Roger would have to ensure that enrollment at the school was above 1,200. Students from the closed downtown school and their families had come to the same conclusion.

Resentful of Roger's tactics and disregard for their wishes, parents began researching alternatives to the board's plans for their students. Over 200 of the downtown students decided to continue their studies online using the government's distance education service, plus home-schooling. Another 150 decided to attend a high school in a different district in a nearby town rather than submit to Roger. Several dozen students decided to continue in the board's independent study program for students experiencing difficulties with a normal school environment. A few families moved away.

On September 1, the 590 Mapleville Secondary students were joined by only 350 downtown students. Roger was furious. He tried to suppress this information by delaying the mandatory reporting he had to file with the Department of Education in October. Mapleville Secondary at 940 students was still below 60-percent enrollment. And he had closed his fully enrolled downtown school. The Department of Education was not pleased. The government wondered why Roger had not closed Mapleville Secondary or at least approached a local community college or the university to develop a co-usage arrangement to use the empty student spaces instead of closing the downtown school. An investigation was launched into the decision-making process that Roger actually used in closing the downtown school.

The transition process of closing the downtown school and merging the students with the Mapleville Secondary students had proved much more costly than anticipated. Then, disgruntled board members began leaking financial information to the media. A senior staffer in the board's finance office took an unexpected early retirement. Another resigned for apparent "health reasons." The merger of the two schools had become a financial quagmire. What should have been an airtight process had turned into a disaster. Trustees were deluged with demands for Roger's resignation. After a protracted period of negotiations, in-camera board hearings, legal jousting, and intervention from the Department of Education, Roger was given a severance package and dismissed the following April.

ANALYSIS

- A dictator can only survive in a closed political system. Roger could not intimidate the supporters of the downtown school who openly opposed his policies. This opposition eventually led to a breakdown of control over his own staff.
- Roger didn't practice the high ethical standards the board expected of its staff and students. He compelled others to support his covert distortions and his lies.
- Roger lost touch with his client base—the students and families of all his schools.
- He tried to pick a winner and hoped that the resulting school-on-school rivalry would cover his devious actions.
- Fear, as a long-term leadership strategy, can only prevail in an atmosphere of respect. Without respect, dissent and eventual resistance is inevitable.
- He alienated a portion of his client base—the downtown students and their supporters—to the point that they sabotaged his plans.

25

Be the Change You Want to See in Others: A Leadership Case History

Ellen was a recently appointed department supervisor in charge of 23 inbound customer service representatives at a large bank. The bank had a very aggressive business culture that rewarded top sales performers lavishly while demanding long hours from poorly paid administrators. Most employees had stretch performance objectives, which encouraged individual excellence, often at the expense of effective teamwork. Ellen was committed to building a cohesive team that would work together instead of always competing for individual performance targets.

While the demanding performance management system at the bank did indeed produce top individual performers, the morale among the customer service reps was poor. Before Ellen joined the department, reps would not share information that might help customers for fear of losing out on the bonuses associated with generating leads for upselling additional products. There wasn't a common understanding of the department's workflow or even which reps were to handle specific product issues. A surprising number of reps didn't know who did what in their department. There was frequent disagreement about reporting timelines for corporate finance. Illnesses, and child-care emergencies (almost all of the reps were women) wreaked havoc on performance statistics for the afflicted individual reps. Colleagues would not share the workload of absent reps because they were not rewarded for doing so. The infrequent department meetings were cold, stilted affairs that Ellen dreaded. These team meetings only generated resentment and rumors instead of building enthusiasm around striving for shared goals. Many of the reps loathed their jobs and hated coming to work; cliques and bitter gossip dominated coffee breaks and the lunch cafeteria. On a business level, the department was not meeting its objectives. Sales were sliding and customer satisfaction scores were mediocre at best.

Ellen was charged with turning this department around. Her operations manager told her to "do whatever you have to do to make this place start producing, particularly the upselling. I'll handle any flack from HR or other business units. Just stick to your budget—no room for special perks." Ellen was committed to not only meeting these performance targets, but also creating a positive work space that was a lot less toxic for her staff. So, she devised a three-point turnaround plan:

- She would start holding weekly department meetings to share targets, progress, but, more importantly, to build a sense of common purpose.
- She started spending a lot more time on the floor with the reps; she helped individuals on difficult calls and made sure that reps who were falling behind in their workload had a chance to catch up.
- She set up a Workflow Analysis Group to ensure that every rep knew his/her job and who to refer specific product issues to when customers complained.

Ellen's first weekly meeting was an absolute disaster. After sharing the previous week's departmental performance numbers, making a few announcements, and recognizing a few individuals for their recent work, she cued up a video clip of a supposedly inspiring motivational speaker on the topic of teamwork. Ellen had purchased this speaker's video out of her own funds. When the final call to action rang out and the music soared at the end of the clip, her audience groaned. Ellen tried to facilitate a discussion on the value of teamwork as a way to debrief the video clip, but the room turned silent as a crypt. Her probing, open-ended questions got little reaction. Eventually, a few individuals came half-heartedly to her rescue, but it was clear to Ellen that she was in the corporate equivalent of a high school study hall on detention duty. After 15 minutes of awkward tedium, she ended the meeting and sent everyone back to their cubicles. Clearly the video clip had missed the mark completely. Ellen wondered if perhaps her team needed to go on an offsite teambuilding retreat and get to know each other better in an unfamiliar context. But, she knew she didn't have the budget for such an event.

Her second strategy of helping out on the floor proved to be more successful. First of all, she learned a lot more about the challenges her people faced, day in and day out. She developed a better sense of who were the power brokers in her department; she knew she had to get these two individuals to come onside with her plan. Without their commitment, she would be continually paddling upstream. But the greatest benefit of her time on the floor each day was that her people saw her helping others. Ellen's predecessor was an aloof number cruncher who only ventured into the cubicles when there was a problem. Suddenly, the reps were experiencing a different, more social style of leadership.

To launch her third strategy, Ellen set up a meeting with the two power brokers in her department. She told them that the workflow of the entire department needed to be mapped out and made more efficient and that she needed their help. This empowering of her unofficial co-leaders was a huge risk. While both were competent, one was a strongly negative leader, and the other was only conditionally positive. Ellen asked for their input on who was to be on the Workflow Analysis Group of five, but reserved the right to make the final decisions herself. This proved to be a wise move on her part as power brokers would have only selected their subservient friends. At first, Ellen attended the workflow meeting to ensure that there was someone in the room who could facilitate the process and help with decision-making tools and shaping the final product.

The weekly departmental meetings continued, albeit, in a shortened format. Ellen did not attempt any more teambuilding initiatives in these meetings. While she intentionally kept these meetings short and focused on immediate business issues, she noticed a gradual shift in tone. The meetings were becoming a bit less chilly. People were starting to socialize a bit more. There was more laughter in the room. What was happening?

Ellen continued to spend a few hours most days on the floor helping individuals. She continued this practice despite pressure from her manager for missing business-unit meetings and falling behind on her reports. Ellen was convinced, however, that one-on-one time with her people was essential for her department at this point in order to create the turnaround her manager wanted.

The Workflow Analysis Group continued to map out the upstream and downstream relationships for every task in the department. While Ellen had to intervene at times in order to support participants who were not highly valued by the power brokers' clique, the meetings began to produce positive results. Her small meeting room became wallpapered with flow charts and Post-it° notes. People from outside the dominant clique began volunteering information to the group.

Ellen continued pushing her turnaround strategy. After four months, she noticed a number of positive changes. Her people were more at ease; individuals who had never talked with each other before started interacting, sometimes on a daily basis. There was more levity in the departmental meetings. When she was out on the floor, Ellen noticed more friendly chatter and fewer stressful outbursts. Her Workflow Analysis Group presented a final report to an audience that included not only the full department, but also a few heads from other business units and Ellen's manager. Clearly this group enjoyed the temporary celebrity that came with making this presentation. But, most of all, the department's upselling numbers had improved for the third consecutive month. Something had changed. Ellen's department was becoming a functioning team.

During a quarterly cross-business unit meeting, an HR consultant asked Ellen to share her teambuilding secrets for turning around her department. "There has been a noticeable change in Ellen's area. Morale is up, absenteeism is down, and I understand that her upselling numbers have improved. Ellen, what teambuilding program did you use?"

Not anticipating the question, Ellen paused, then asked for a clarification, "What teambuilding program are you referring to? The only thing I did was show a motivational video that went over like a lead balloon. They hated it! All I have done is spend a lot more time on the floor helping individuals, and, with the help of a working group of my reps, develop a complete picture of our department's workflow. That's all." The HR consultant said, "But surely you must have done something else; the whole team is a lot more cheerful and I hardly get any calls for EAP [employee assistance program] counseling these days. I heard the other day that they run your weekly department meetings now, and you just attend as an observer. That's a huge change!"

Ellen shrugged, then smiled, "Beginner's luck, I guess."

What did Ellen do right? How did her turnaround happen? Was it all just due to a supersized helping of "beginner's luck"?

ANALYSIS

- Ellen did a few things wrong, but a lot more things right, hence her success. She made some mistakes, but persevered by showing that she trusted her people.
- She quickly realized that her team would not respond to pep talks or inspirational teambuilding activities. She knew such activities would just breed more cynicism, so she stopped trying.
- More than anything she *said*, the customer service reps on the floor began noticing what Ellen *did* almost every day. She worked with them. She helped individual reps get their work done on time. She modeled the teamwork she wanted to see on the floor. She didn't lecture or cajole. She modeled the cooperation she wanted to see. She just did it as opposed to talking about it. Ellen demonstrated the truth of the old maxim: "Actions speak louder than words."
- Again, through her actions, Ellen demonstrated by setting up her Workflow Analysis Group that she respected the abilities and the experience of her people. She trusted them to sort out what they needed to do in order to create a positive work environment.
- Ellen realized that effective teamwork is not an objective, but an indicator of shared agreement and clarity about roles, goals, values, and work processes.

Index

A

academia, 27–28
acknowledgment of issues, 72, 73
agenda
    controlling in interviews, 63, 65
    giving feedback, 105
    meetings, 99, 100
agreement, coaches, 86, 87
alignment, coaches, 86, 87
allies into enemies, 45–48
amateur status, 113
analysis
    "dictatorship," 119
    positive changes, 125
appreciation, 72, 73
asking better questions, 13–16
assessment tools, *see also* Innovation tools
    building stronger teams, 38–40
    strengths-based organization, 42
associations, forcing, 91, 92
assumptions
    asking better questions, 15, 16
    switching, 91
    working together as teams, 38
attitude, *see* Chemistry
autonomy, 50, 52
avoidance, communication, 85

B

Baby Boomers, 97, *see also*
        Intergenerational conflicts
balanced life, 29, 30
balanced picture, 62
"been there, done that," 37
behavior, *see also* Capability
    case history analysis, 119, 125
    feedback, 105
    importance, 5, 7
believing the worst of others
    adapting coaching, 18–19
    teams, 38, 48

benefits
    conflict, 7
    interviews, 61
    selling enemies on, 48
    working together as team, 38
Best Opportunity idea, 90–91, 92
biases
    better business decisions, 79, 81
    preparing for interviews, 63, 65
blame, 105
Boomers (Baby), 97, *see also*
        Intergenerational conflicts
Bottom Line, as measure, 28
boundaries, 22, 25
brain downtime, 10, 12
brainstorming, 90–91, 92
brand, knowledge of, 54
Britain's Naval Intelligence Division, 3
Buckingham, Marcus, 41–42
Buffet, Warren, 79
buttons, pressing, 23–24, 25
buy-in establishment, 100

C

candidates, evaluation of, 62
capability
    delegating work, 57–60
    interviews, 61–62, 65, 66
    matrix, 91, 92
career, *see* Personal career
carefulness, 5
car engine analogy, 33
Carl (example), 45–48
carrot-and-stick motivation model, 49
causes, *see* SCCIO analysis
cell phones example, 77
certification, coaches, 19
challenge statement, 90, 92
changes
    accelerated, 18
    being what's wanted, 5
    positive, 75–78

About the Author

William G. (Bill) Templeman is a writer, instructor, designer, and coach with 20 years of experience working with corporate, public sector and nonprofit clients. He has a strong background in experiential education (action learning), training program design, and process facilitation. He holds a B.A. in psychology from Concordia University in Montreal and an M.A. in English from the University of Toronto.

Templeman has worked with clients across Canada and the United States. He has written, designed, and delivered a wide range of training programs to employee and management groups including

- Team effectiveness
- Leadership development
- Managing change
- Career transition
- Sales/client service
- Coaching

Prior to starting his own business in 1993, Templeman worked for five years as an internal management development consultant at Royal Trust in Toronto. He currently runs his own consulting practice in training, coaching, and business communication. In addition to his consulting and writing, Templeman also delivers communications and career search courses at Fleming College in Peterborough, Ontario. He also works with clients as a career counselor for Lee Hecht Harrison throughout eastern Ontario.

Templeman is also a former wilderness program course director and has an extensive background in experiential team building and leadership development with both employee and management teams. His writing on business, politics, and education has appeared in national magazines, journals, Web sites, and newspapers. He can be reached at: bill@edgeworkonline.com